'*Seven Ethics Against Capitalism* sharply reveals the multiple crises being generated by the capitalist mode of production – from climate breakdown, to inequality, to the erosion of democracy – and how impossible it would be to fix any of these problems without a radical transformation in the way we organize society. Mould convincingly argues that values such as solidarity, stewardship and radical love must be at the heart of this new vision for the world, as well as the movements aiming to bring it into being.'

Grace Blakeley, author of *Stolen: How to Save the World from Financialisation*

'In times when everything from nature to desire is being privatized, to shift our attention towards planetary commons is an essential ethical duty. Oli Mould's timely *Seven Ethics Against Capitalism* is an urgent and passionate call not only to deeply rethink our dire present but to create the conditions for our common future beyond capitalism.'

Srećko Horvat, author of *After the Apocalypse*

'A conceptual toolkit and survival guide for outliving capitalism. Through an original, compelling and readable account of the commons, Mould distils a set of ethical dispositions for building a more convivial and democratic future.'

David Madden, London School of Economics and Political Science

Seven Ethics Against Capitalism

Seven Ethics Against Capitalism

Towards a Planetary Commons

Oli Mould

polity

Copyright © Oli Mould 2021

The right of Oli Mould to be identified as Author of this Work has been asserted in accordance with the UK Copyright, Designs and Patents Act 1988.

First published in 2021 by Polity Press

Polity Press
65 Bridge Street
Cambridge CB2 1UR, UK

Polity Press
101 Station Landing
Suite 300
Medford, MA 02155, USA

All rights reserved. Except for the quotation of short passages for the purpose of criticism and review, no part of this publication may be reproduced, stored in a retrieval system or transmitted, in any form or by any means, electronic, mechanical, photocopying, recording or otherwise, without the prior permission of the publisher.

ISBN-13: 978-1-5095-4595-7
ISBN-13: 978-1-5095-4596-4 (pb)

A catalogue record for this book is available from the British Library.

Library of Congress Cataloging-in-Publication Data

Names: Mould, Oliver, author.
Title: Seven ethics against capitalism : towards a planetary commons / Oli Mould.
Description: Cambridge, UK ; Medford, MA : Polity Press, 2021. | Includes bibliographical references and index. | Summary: "From 'failure' to 'slowness', how to throw off the shackles of capitalism"-- Provided by publisher.
Identifiers: LCCN 2020052810 (print) | LCCN 2020052811 (ebook) | ISBN 9781509545957 (hardback) | ISBN 9781509545964 (paperback) | ISBN 9781509545971 (epub)
Subjects: LCSH: Capitalism--Moral and ethical aspects.
Classification: LCC HB501 .M7629 2021 (print) | LCC HB501 (ebook) | DDC 174/.4--dc23
LC record available at https://lccn.loc.gov/2020052810
LC ebook record available at https://lccn.loc.gov/2020052811

Typeset in 11 on 14 pt Sabon by
Servis Filmsetting Ltd, Stockport, Cheshire
Printed and bound in Great Britain by Short Run Press

The publisher has used its best endeavours to ensure that the URLs for external websites referred to in this book are correct and active at the time of going to press. However, the publisher has no responsibility for the websites and can make no guarantee that a site will remain live or that the content is or will remain appropriate.

Every effort has been made to trace all copyright holders, but if any have been overlooked the publisher will be pleased to include any necessary credits in any subsequent reprint or edition.

For further information on Polity, visit our website: politybooks.com

Contents

Acknowledgements vii

Introduction 1

Ethic 1: Mutualism 33
Ethic 2: Transmaterialism 57
Ethic 3: Minoritarianism 77
Ethic 4: Decodification 97
Ethic 5: Slowness 120
Ethic 6: Failure 140
Ethic 7: Love 158
Conclusion: The State of the Commons 177

Notes 191
Index 206

Acknowledgements

It is with a certain amount of trepidation that I attempt to 'acknowledge' all the people that have written this book with me. Indeed, if you are reading this afterwards, it is clear that having my name on the front cover sits rather uncomfortably with the goals of the book – for there are so many other people and things whose agency flows through and over the pages.

The first group of people I need to mention are my wonderful colleagues at Royal Holloway and beyond in the wider academic world. The university sector in the UK is under attack by capitalist forces, and it is only the camaraderie and solidarity between the staff that seem to be stopping these forces from destroying it completely. My corridor neighbour Innes Keighren has always been a sounding board and I have lost count of the number of times his office has become a group therapy gathering. Others from Royal Holloway Towers and beyond, including Alasdair Pinkerton, Sofie Narbed, Phil Brown, Mike Dolton, Katie Willis, Simon Springer, Katherine Brickell, Max Haiven, Mel Nowicki, Ella Harris, Cecilie Sachs Olsen, Thomas Dekeyser, Sasha Englemann, Phil Crang, Pete Adey and Rachael Squire, have all been there

Acknowledgements

to bounce ideas off and listen to me rant. And my PhD cohort – Emily, Ed, Megan, Rhys, Jack, Angela and Will – you are all absolute saints to be supervised by me and having to listen to my incoherent yet constant ramblings. As it turns out, though, some of these have informed the thinking in these chapters after your sage advice. I must also acknowledge those on the picket line during the UCU strikes of 2019 and 2020, Dan Elphink in particular; his folk guitar and protest songs were so warming it was easy to forget the sub-zero temperatures in the air around us and in the corridors of power. And then there are my brilliant students, without whom many of the thoughts that contribute to this book would have remained unsaid.

I must also thank the editor of this book, Jonathan Skerrett, who has was bold enough to take a punt on the book, and has guided it through choppy waters and some even choppier reviews. He has been immense throughout and I wholeheartedly thank him for all his hard work. I am also extremely grateful to the anonymous reviewers who offered comradely and critically constructive advice on the theories of this book; indeed some of the ethics have changed in response to their soaring intellects and so whoever you are, you are co-authors.

I also want to acknowledge my dear friend and pastor Mark Woodward. His sermons on the radical love of our saviour Jesus Christ have been inspirational to me, and he has always been there to listen to my sometimes wacky but always passionately argued (!) theological contortions. His work is particularly evident in the Ethic of love, so thank you, Mark; keep on keeping on! Also Will Lowries, John Wills and Jonny Hopper, you have all been inspirations in my spiritual outlook on life, and will no doubt recognize many conversations we've had in the pages of this book. There is also the not-so-insignificant matter of thanking

Acknowledgements

Nihal Arthanayake, Mark and Neil Pearce and Sam Fender – they'll know why and just how much it meant at the time and what it still means now.

But it is to my all-female family that I owe my greatest debt. My mother is always there to support, pray and look out for me (and occasionally point out my grammatical errors on Twitter). My incandescent wife Sarah, a front-line general practitioner, is an inspiration to me. In her dedication and selfless loving compassion for everyone she meets, she is a personification of the kind of social world I want to see flourish. She is a beautiful healer of broken bodies, hearts, souls and minds; I owe her everything. And my children Penny and Jessica have radically shifted my worldview for the better and continue to be an inspiration in everything I do. I love you all with everything I am (even the latest addition, Ginger the hyperactive dog).

Finally, I want to dedicate this book to my late father, Graham, who died of mesothelioma while I was writing this book. He was a devoted father and brilliantly patient man, who supported me (and my two brothers) in whatever it was we wanted to do with love, compassion and an unshakable faith. Even though he had lived a full life, he was still taken from this world far too soon and has left a gaping wound in the lives of those who were lucky enough to call him a friend. He is the reason for me being me; I cannot thank or love him enough.

Introduction

Capitalism isn't working. Over the course of the twentieth century it colonized almost every nation of the globe. Yet, in the first two decades of the twenty-first century, it has hastily ushered in the emergence of growing climate catastrophe on a planetary scale. There is little point in trying to tweak the way capitalism works to be more ecologically sustainable, because its underlying and foundational principle of privatizing the means of production entails the extraction of natural resources to an ever-deepening scale in the all-consuming pursuit of 'growth'. Capitalism cannot be fixed. The half a millennium or so of rampant imperialist mercantilism, which mutated into a nefarious neoliberal global capitalism and now has morphed into a dangerously fascistic form of nationalistic wealth generation, has proved beyond any reasonable doubt that capitalism does irrevocable damage to the planet, to the climate, to biodiversity and to us as a species.

What is more, all the benefits that supposedly flow from capitalism – creativity, liberty, morality, enlightenment, equality, democracy, wealth and the progress of civilization – are now drying up, and in some cases reversing completely. And more recently, this has been exacerbated. Because the coronavirus pandemic that swept the globe in 2020 rocked capitalism to its very foundations; and it has shown just how much we depend on each other, not capital, for survival. The response to the spread of the virus and the need to keep people 'locked down' saw the revival of state-level quasi-socialism on a level barely seen in a generation. There were some of the largest financial bailouts by governments the world has ever seen, to industries and workers. Once bastions of capitalist society such as the US and the UK rapidly implemented policies that were the mainstay of socialist demands, such as rather thinly veiled versions of universal basic income, student debt cancellation, free public services and, of course, the pedestalling of socialized healthcare. The nuances of these are still being implemented, and while a vaccine has been found and the virus will be managed, its impact upon the future of national institutions and indeed society more broadly will be felt far into the future. Because of the impact of the coronavirus, and the now increasing need for governments to act in similar ways to combat the inevitably far bigger crisis of global climate catastrophe, the weaknesses, inefficiencies, inequalities and injustices of capitalism have been vividly exposed.

Yet despite this, the advocates of a capitalist way of life continue to preach that the only way to achieve progress and a better and greener world is to blindly continue along the same path of destruction we have travelled on for so many years. But they are wrong. To tackle the global problems of the future starting with the present climate catastrophe, capitalism needs to be replaced with something else entirely.

But how can this be done? What should replace it? History is littered with revolutionary events, when the oppressed rise up and overthrow their capitalist masters and attempt

Introduction

to install a fairer form of society. But from small-scale, local changes to generation-long episodes of state-led communism, they have all – to a greater or lesser degree – fallen foul of the lure of capitalist dogma that decrees 'there is no alternative'. This is because in attempting to 'scale up' anti-capitalist societies, these revolutionary events have – in one way or another – started to (and in some cases, completely) mirror the injustices of capitalism by invoking the same kind of power imbalances, authoritarianism and inequality, just with a different political economic hue. Their anti-capitalist fundamentals have been lost.

But there is one societal ideology that has remained constant throughout these episodes. From human prehistory, throughout capitalism's growth, and all those failed revolutions, the very real ideology of the *commons* has remained. Now, it is an idea whose time has come. But in order for it to aid in the reconstitution of our planet and the healing it requires, we need a *planetary* commons. This is the coming together of *all* peoples and resources in the world into a *planetary* (not global, or international) mode of socio-economic organization that recognizes our material, cultural and psychological intimacy with the planet we inhabit and the human, nonhuman and intangible resources it offers. Planetary thinking embraces the differences of and in the world, and as feminist scholar Gayatri Chakravorty Spivak has argued, it resists the image of the 'globe' or globalization as a false totality.[1] Practically, then, the planetary commons is a mode of organizing communities, nations and societies that foregrounds the very characteristics that capitalism defenestrates. Solidarity, stewardship, protecting the vulnerable, slowness, and even love; these are some of the ethical ways of being that capitalism diminishes, yet are vital if a planetary commons is to come into view.

Seven Ethics Against Capitalism

Sometimes hidden, the traditional view of the commons has historically – either physically or ideologically – always been a means to subvert, resist and critique the prevailing order of social organization (be that feudalism, fundamentalism, a dictatorship or, today, capitalism). The commons has provided people and communities throughout history with a mode of existence within the cracks of hegemonic societal systems that we are living under. Today, within the cracks of capitalism, a common world is flourishing.[2] The commons, as an ideology of human community, has existed in, through and outside of the prevailing order of society for millennia. It has provided societal organizations that are, on the whole, not only ecologically sustainable, but more equitable and just. The commons doesn't need the creation and exchange of capital to thrive; it only requires the willingness of those who believe in it to succeed.

However, caution is clearly required because the predatory growth of capitalism in the twenty-first century feeds off those forms of life that exist 'outside' of it. Appropriating anti-capitalist motifs,[3] accumulating by dispossessing,[4] and violently enclosing land, societies and ideologies that are not conforming to the mantra of profit-maximization, capitalism thrives off those people, places and experiences that critique it. And via its leading edge of marketing, public relations, advertising and the vernacular of 'creativity', capital is created out of the eventual privatization of that which was once held in common. Land, nature, housing, knowledge and even creativity itself have all been wrenched out of common ownership and been carved up and profited from by frontier capitalists. And that which is still common (e.g. the internet, the air we breathe and, now, outer space) is being targeted for privatization and subsequent commercialization.

Therefore to grow the commons to a point at which it

Introduction

is a viable social alternative requires protecting it from enclosure by contemporary forms of capitalism. It requires an active *anti*-capitalism that is also simultaneously a form of growing the commons, something that political geographers Gibson-Graham call common*ing*.[5] Commoning as a practice by *some* creates more commons as a resource for us *all* to benefit from. Despite the voracity of capitalism's enclosure, there are examples of communities building a commons that is not only resisting this process, but also expanding the more sustainable, just and equitable social organization it creates *back* into the capitalist world for everyone to share. For example, there are inner-city squats that have resisted enclosure for decades and are now beginning to inform how cities are being built beyond the pure pursuit of profit; community gardening groups that have influenced urban agricultural practice to be more ecologically sustainable; refugee activist groups that have made spaces for people otherwise trafficked and brutally oppressed; climate justice movements that transformed city centres into enclaves of play, theatre and protest and are now shaping national and international policy on climate change; factory workers who have forcibly taken over the management structure to make it more equitable for all workers; and, in the wake of the coronavirus, mutual aid networks that have sprung up all over the world to help deliver food to the isolated, care for the sick, and educate and entertain locked-down children.

These are already-existing (and spreading) examples of the anti-capitalist commons that show how alternative ways of organizing our economies and societies are possible beyond the injustices of capitalism. They point tantalizingly towards a future beyond the environmental and societal injustices that we currently endure. They showcase the kinds of practices, behaviours and mindsets that have not only resisted capitalism, but built fairer worlds. But only a radical emancipation and diffusion of those already-existing commoning practices into a powerful collectivized force can see it viably resist capitalism. Before we can even begin to think about what structures, institutions, policies, governments and cities we need to build, there needs to be a radical change in the *ethical* position of our societies to reflect the emancipatory potential of the planetary commons. Wrestling back, maintaining and then spreading the commons away from a predatory capitalism requires ever more physical, virtual and emotional resources from those people invested in the commons' survival (which, if we are to avoid the omnicide that a capitalist realism is marching us towards, will need to be everyone). In short, these resources need to be harnessed, to create an ethical commitment to realizing a *planetary* commons before it is too late.

This book therefore proposes a set of seven ethics that are gleaned from the already-existing commons. Individually, they can be seen as characteristics of the commons that are in direct opposition to the deleteriousness of capitalism. They are ethics *against* capitalism. But together, they can act as a mode of understanding the broader movement of commoning, and how it has the potential to resist and undo the deleterious effects of the current prevailing world order. What they are not is a static blueprint for action, a hegemonic view of a new planetary order that will only replace one form of ideological colonialism of the world with another. Indeed, scholars have argued that many of history's most barbaric colonial acts, not least the destruction of indigenous Americans by European 'pioneers', are tied up with the imposition of 'common land' for the settlers.[6] Instead, commoning is a way of being-in-the-world that disrupts the smooth functioning of

Introduction

the capitalist status quo and its planetary violence on all peoples. So together, these seven ethics are a call to rethink and re-engage with the planet in more just, equitable and ecologically sustainable ways that will safeguard our future-in-common. I will outline in detail what I mean by ethics, but first, what do I mean when I say the commons? And how can they be planetary?

The commons

There is no shortage of definitions and articulations of what is fundamentally a very elusive concept. The term 'common' refers perhaps to banality or the mundane, maybe a shared interest between friends, or even a derogatory slur upon a particular class of people. As easily dismissed as these can be as part of the quotidian vernacular, there is an underlying sense even with these uses that we can experience a shared existence that transcends a superficial individuality. Beyond that, though, 'the commons' becomes a slippery concept. But such elusiveness is a symptom of its vitality in human existence; knowing what the commons is and crucially *how* to enliven it is as deep a human trait as can be thought of. We are social creatures, we all descend from the same primordial soup, and we live in and share the *common*wealth that this planet affords us.

Fundamentally, the commons is that which we build by being together. More than a natural resource – a forest, a lake, a field – the commons is the community that builds up around and beyond it, the society it creates and the continual act of democratizing access and sharing the gifts of that resource to those who need it most. Building on the work of anthropologist Stephen Gudeman and geographers J. K. Gibson-Graham, the commons can be thought of less as a unitary or singular protected 'natural' resource (such as a rainforest, a pasture or an irrigation system, which are traditionally thought of as 'common' resources in institutional narratives) and more as a dialogical creation between a resource and the community it brings into existence. In other words, to realize its emancipatory potential from capitalism's enclosure, the ontology of the commons requires a deeper understanding of its 'lived' component – something that comes from an interaction between the place and the community that relates to it. This conceptualization of the commons therefore differentiates it from those seen in more institutional and global forms, namely the Bretton Woods institutions such as the United Nations, the International Monetary Fund or the World Bank (but also including national-level interests such as foreign aid departments). These tend to see the commons as a static piece of land or natural resource that falls outside the jurisdiction of national governments or private interests; something to be guarded, with access limited to a deserving few.

Gudeman and Gibson-Graham argue against this. For them the commons is not a physical resource that abides by some regulatory framework that is imposed from above. They argue against this 'top-down' institutional view of the commons. They do acknowledge the importance of safeguarding the material wealth of the commons, but without theorizing the resources as being *of* the community, the commons will continue to be threatened with capitalist enclosure. This is because the commons will still be beholden to the same global political-economic logics that dictate the global institutions and national governments, that is, market interests ultimately trumping those of the indigenous communities. It's just for the supra-national institutions these logics have agreed to collaborate via regulation to administer the scarcity of the resource. It is of course laudable to protect a resource from overuse (and indeed has helped protect rainforests around the world from deforestation, oceans from overfishing and pastures from overgrazing) but it is not a functional mode of *diffusing* the commons throughout society so as to resist and replace capitalism, because it is ultimately beholden to market logics, however steeped in social responsibility they may be at the time. Indeed, the institutions that govern them will often restrict local indigenous communities from accessing the common resource, designating it instead as a protected area and assuming a stewardship role, dividing up the resource as *the institutions* see fit, rather than collaborating with local knowledge; it is a form of colonial commons.

This is not a concept of the commons that we need today. Instead, any commons does not exist until a resource is overlaid with a community of people (and things) that freely access it. Gudeman argues that 'taking away the commons destroys community, and destroying a complex of relationships demolishes a commons'.[7] Seeing the commons in this way redefines both the commons *and* community. As the feminist scholar and researcher of the commons Silvia Federici argues:

> 'Community' has to be intended not as a gated reality, a grouping of people joined by exclusive interests separating them from others, as with communities formed on the basis of religion or ethnicity, but rather as a quality of relations, a principle of cooperation and of responsibility to each other and to the earth, the forests, the seas, the animals.[8]

An example of this conceptualization of the commons often cited is the Van Panchayats in India, an indigenous community-based forest management system that came about through protests in the 1920s against what the community saw as the mismanagement of the forests by British Imperial rule. The community broke away from the state-led 'Joint Forest Management' system, which they saw as ineffective in stopping deforestation and the decline in local biodiversity. For a century, and at much lower cost than this national scheme, the local communities have continued to live in and off the forest as an integral part of their daily activities, all the while maintaining biodiversity levels and managing de- and reforestation themselves.[9]

Another example that extends this idea into the sociopolitical realm is that of Cherán in Mexico, a town that was ravaged by illegal loggers and with a corrupt local government that turned a blind eye. The locals ran them both out of town and have never let them back in. That was in 2011, and today, the town does not take part in local or presidential elections, has its own community-led security force – *ronda* – and governs via a group randomly selected every three years.

There are many other examples that will be alluded to throughout this book that point towards how the commons is more than a specific natural resource. It is important to note, however, that this conceptualization of the commons is not entirely new. If we delve into the etymological history, there are glimpses of this kind of planetary commons evident throughout its long and complicated epistemological construction. It has spiritual, material, political, economic and cultural underpinnings that, if teased out, can help us to affirm the kind of commons that a planetary reading of it entails. So a brief and potted history of the commons is worth outlining.

Introduction

A history of the commons

As mentioned previously, the commons is a nebulous concept, and so pinning down a history is a perilous task. History itself is a hegemonic project of enclosure, with those events, theories, ideologies and philosophies that were recorded given credence over those that were not. As such, analysing a history of the commons with the material available will inevitably err, because it relies on that which is written down (and accessible to me as an English-speaking, lowly academic researcher). So it is vital to recognize from the outset that various articulations of the commons – from a spiritual, philosophical and natural standpoint – have existed as long as humans have. From theories of property laws in Mesopotamia,[10] the ancient Egyptians' belief in the unifying force of Ma'at, the Andean goddess Pachamama, Confucianism and Taoism in ancient China and the Druids in ancient Britain, to animism among indigenous peoples, there are ancient and non-Western narratives of the physical and spiritual commons that still exist, and thrive, today. However, to grasp how the commons has developed into an ideology that exists alongside, but with the potential to resist, contemporary forms of capitalism, it is pertinent to start a history of the commons, for this researcher at any rate, at the genesis of that capitalism, namely in ancient Greece.

Heraclitus (who died c. 475 BCE) was an Athenian philosopher, and insisted that we as humans, in order to become civilized and progress as a species, must 'follow the common' – the common of the *logos*. The logos, for Heraclitus, was a philosophical concept. It was not pure order, logic or reason (as is sometimes inferred from the etymological lineage); he used it far more esoterically to denote the cosmic 'existence' beyond our understanding. In some of his quotes it is 'the mind of God' but in others not a *super*natural force at all; instead it is the 'language of nature'. The logos was for Heraclitus a common experience for everyone. In his quoted sayings, he insisted on the metaphor that for those who are awake, there is only one world in common, but those who sleep withdraw into a private, self-interested world. We must therefore not act and speak as though we are asleep, but adhere to the common logos, forgoing private lives.

Moreover, the logos is unifying because it incorporates paradoxes and opposites, such as the 'ways upwards and downwards are one and the same', and 'the beginning and the end are common'. Moreover, Heraclitus' most famous quote is 'you can never step into the same river twice'. In saying this, he was indicating that everything flows from and in the common; there is no stasis or fixity, as everything is constantly in flux; but it is the same river that flows; it is the same logos that flows. For Heraclitus, then, there is unity in the world – as each opposite cannot exist without the other – but it is a unity that flows, is never static and always changes. 'Following the common' for Heraclitus was the way to enlightenment, peace and self-control. Being 'asleep' and deviating from the common logos was to be ignorant of the truth.

Fast forward two millennia or so and the indigenous populations of the Americas are being systematically enclosed, marginalized and murdered by settler colonialists from Europe. The genocide of the Native Americans by various European monarchs in the fifteenth and sixteenth centuries is entangled with the realization of 'the commons', albeit as a precursor to private property rights. The 'native' indigenous person was seen by the Enlightenment scholars of the time as the Noble Savage 'at one' with

Introduction

nature, outside of modernity and as such part of the commonwealth of the land.[11] The mutual respect shown by the native Americans and the ontological equivalence with themselves that they afforded to the land and the animals were very much part of their ancient indigenous spirituality; but very much at odds with the European mindset of seeing the world as a resource ripe for primitive accumulation. As such, the natives became simply another natural resource for the Europeans to commandeer and carve up among their fellow settlers, or indeed to ship back to Europe as slaves.

But even within Europe itself, the commons was present, though under attack from enclosure. The Diggers were a group of radical Protestants who (in the wake of the First English Civil War) believed that humans are implicitly connected with nature, and the ownership of land by individuals was unjust, immoral and illiberal. Their main protagonist, Gerrard Winstanley, wrote in 1652, 'true freedom lies where a man [sic] receives his nourishment and preservation, and that is in the use of the earth'.[12] The Diggers set up communes across England, the most prominent being at St Georges Hill in Surrey (now, ironically, one of the most expensive privatized and gated communities in the entire world). Although their communes were eventually dismantled, they went on to form other groups, notably the Levellers, who carried forward the idea of the 'common' as an alternative to private land ownership. They championed a 'commonwealth', a land that could produce an abundance of resources free from what they saw as the tyrannical rule of the monarchy; a common *wealth* for everyone. They were, of course, battling against a growing belief in self-interest as the driving force of liberty, one that reaped massive rewards for the aristocratic and mercantile elite. And so the Levellers were quashed before they could mobilize political and resistive momentum (we will return to the Diggers in Ethic 2).

Their ideological, Heraclitian stance on the commons, though, remained, and can be exemplified in many struggles across the world in the subsequent centuries. The most notable from a political perspective was in the Paris Commune in 1871, when thousands of workers, servants, refugees and middle-class Parisians blockaded themselves in the city in response to the violence of the French government. The influence of the Paris Commune on future radical political movements cannot be overstated, but what is important here is their commitment to commonality, to the rejection of individualism and the violent nationalism it entailed. Indeed, one of the main protagonists of the Commune, Elisée Reclus, argued that 'Everywhere the word "commune" was understood in the largest sense, as referring to a new humanity, made up of free equal companions, oblivious to the existence of old boundaries, helping each other in peace from one end of the world to the other.'[13]

The Communards' notion of the commons led to them creating a makeshift society that lasted for only seventy-two days, but one that focused on shared living and a distinct rejection of self-interest. The bloody end of the Commune at the hands of the French army represents the lengths to which the hegemony of systems of empire, colonialism and the state will go to assert its own version of progress. But the Commune also showed that in just seventy-two days, an actually existing commons was created that still influences political movements and scholarly debate today (we will revisit the importance of the Paris Commune in Ethic 5).[14]

More recently, the concept of the commons has been used to articulate an international common resource, most

Introduction

notably by the economist Elinor Ostrom, who published *Governing the Commons* in 1990.[15] She articulated the already-existing ways in which indigenous communities were effectively and efficiently managing commonly shared resources such as water, forests and grazing land. She was responding to the so-called 'tragedy of the commons', put forward by Garrett Hardin in 1968, who argued that the common use of a resource would lead to its ultimate depletion because of the inherent self-interest of certain individuals.[16] For Hardin, private ownership was the only way to secure the future of that resource. But Ostrom's research saw that many people were rejecting this idea, and she put forward a set of principles that, if adhered to, can sustain a common resource and not lead to its ultimate depletion. She won the Nobel Prize in Economics in 2009, and many of her ideas are now implemented by the World Bank and their like to govern precious natural resources such as the Amazon rainforest. However, such institutionalization of Ostrom's principles has led to privatization by another route. Silvia Federici has argued that the World Bank (and other supra-national Bretton Woods institutions such as the United Nations and the World Trade Organization) have commandeered important natural resources and merely put them under the control of states (which have been largely hollowed out by corporate interests). And under the guise of 'protecting biodiversity', access is limited to certain privileged companies, dignitaries, tourists and investors, all while indigenous communities continue to be expelled.

Global material resources are one thing, common global cultures and ideologies are another. There is very little or no cost of reproduction to a commonly consumed radio broadcast, film, creative idea or other cultural product; once it is made, it can be consumed without cost again and again by more and more people, potentially ad infinitum (unlike food or energy). Political theorists Hardt and Negri argue that the commons, enacted by an internationalist 'multitude' of people resistive to globalized capitalism, can also be 'the languages we create, the social practices we establish, [and] the modes of sociality that define our relationships'.[17] Yet even this more Heraclitian view of the commons is being enclosed by contemporary techniques of capitalist accumulation. Intellectual property rights (and the aggressive legal defence thereof), the hyper-individualization of everyday life by personal technologies and social media, and the quantification of everything (so as to be more amenable to markets) are just some of the processes that are enclosing 'common' shared global socio-cultural experiences. Cultural artistic forms such as music, film and TV that have been collectively experienced and enjoyed are now being deliberately produced to appeal to algorithmically created playlists, accessible on personalized media rather than speaking to social issues more broadly. Marginalized collective (sub) cultures are being appropriated for commercial gain. The very relationships we have with our friends and family are being filtered through technological interfaces that optimize advertising revenue. All this stifles any sense of a common collective culture as an alternative to a digital, individually tailored and highly commercial form of capitalism.

So all of these articulations of the commons[18] – a spiritual and philosophical collective humanity, a commonwealth of land, a political resistance of radical equality, an economic rationale and a cultural collective – are vital to understanding what it is that we create by simply being together. But it is the interplay between the common resource that is created and the community it creates and sustains that is

Introduction

the cornerstone of any further conceptualization of the commons.

The planetary commons

Yet today, as we reel in the wake of a deadly coronavirus and attempt to learn lessons as to how to survive future pandemics better, as well as continuing to face climate catastrophe, realizing the commons on a planetary scale is more crucial than ever. In building the commons as a dialectic between community and a common resource, the recognition that the spiritual, land-based, political, institutional and economic commons that we create are extensions of the planet we inhabit is vital.

The mixing of the planet's resources with our labour for millennia has created a world that we cannot extricate ourselves from. And as we have depleted those resources to critical levels, so too have we depleted ourselves. Capitalism, particularly the neoliberal-soaked versions of it that have produced a re-emergence of governmental fascism,[19] mental health epidemics,[20] violent borders[21] and a chronic inability to deal with pandemics, is itself a pathogenic symptom of our disconnection with the world. The more we drive an ideological wedge between who we are as a species and what the planet is as a living resource, the more damaging capitalism has become.

Adding the prefix 'planetary' to 'the commons' augments the conceptualization of the commons as a community–resource dialectic in two related ways. First, it extends the co-creation of both the commons and community to include the planetary resources that we have been intermingling with ever since we began using tools as a species. The land, subterranean resources, forests, fauna and flora, the atmosphere, our bodies and near space: the materiality of the planet (and beyond) is a complex mingling of things and people being created, destroyed and recreated continually. But more than that, the intangible parts of our collective life – culture, society, economies, communities – are products of how we have used the material world around us. Even the 'virtual' commons of the internet, globalized culture and the zeitgeist itself uses materials to sustain itself; the cables, servers, smart devices, geostationary satellites and raw materials that they are made from are just as vital to the creation of the digital commons as the creative and cultural content that populates it.

Furthermore, a *planetary* mode of organization recognizes our material and psychological intimacy with the planet as *Gaia*.[22] As the philosopher Bruno Latour has argued, we need to 'rematerialize our belonging to the world'.[23] Within this process there is the necessity to resist totalizing narratives that reduce the heterogeneity of the world's population into a single homogeneous entity. As Latour (among others) has continually stressed, the nature/culture divide is a false one, and attempts by culture to curb and control nature are at the root of capitalist-induced climate catastrophe. A planetary commons rejects this divide and calls for a 'reterrestrializing' of our existence in the world.[24]

Second and relatedly, in their conceptualization of the 'planetary turn', the scholars Amy Elias and Christian Moraru have argued that globalization is a totalizing and homogenizing force, one that is suspicious of difference as an inefficiency in the smooth functioning of global capital across the many parts of our world (i.e. the Bretton Woods institutions and their allocation of the 'global commons'). Globalization is the creation and maintenance of the global scale that contains the flows of capital and the elite, at

Introduction

the expense of the nuances of the local. Globalization is a homogeneous force that seeks to annihilate difference. Instead, Elias and Moraru talk of planetarity or 'worlding' as something that focuses on relationality and, crucially, ethics. They argue: 'Planetarity is configured – artistically, philosophically, and intellectually – from a different angle and goes in another direction [from globalization]. It represents a transcultural phenomenon whose economical and political underpinnings cannot be ignored but whose preeminent thrust is *ethical*.'[25]

Planetarity is less thinking the world as the same than celebrating its difference. It is a rejection of the powerful forces that seek to homogenize the world into an abstract consumption product so as to improve the bottom line. Instead, being 'planetary' widens our aesthetic and ideological gaze, and views the world as a multiplicity of cultures, people, places and things, all held together in balance, against a capitalism that is very much *im*balanced. A planetary commons, then, is not one that is global (that would be to the detriment of the local), nor is it international (that would be to fall back on existing geopolitical structures that continue to fail us). Hence, configuring the commons as planetary acknowledges their infectious and contagious characteristics and highlights how they spread to those realms of social life that have been ravaged by capitalism.

On a practical level, this requires *safeguarding* and *protecting* what we have created in common throughout the millennia (and continue to create), but not in the way that is currently done via institutional forms. To shift our focus on the commons to a planetary conceptualization requires an understanding of not only what it is that we create by being together, but how an empirically new form of society can be brought into being that realizes a common world that can be safeguarded from the violence of capitalism. So, more than affirming a simple 'stewardship' of the world's resources (which can often slip into neo-colonial contexts that order society around those who qualify to be stewards and those who do not), a planetary commons denotes a far more intimate connection with those resources we seek to protect.

But in the midst of a powerful, all-pervasive enclosure by capitalism, how is the ideology of a planetary commons to survive? Is the idea of the commons forever to be marginalized? How can the spirit of Heraclitus, the materiality of the Diggers, the political imaginary of the Paris Commune, the economic rationality of Ostrom, the shared cultural internationalism of Hardt and Negri's multitude thrive?

The answer that this book propounds is to rekindle an ethics of the commons and reconceptualize it as not just a potential enclave of resistive anti-capitalism (which of course is important), but as more: as a creative, and infectious force of planetary *commoning* independent of capital. As Gibson-Graham have argued, if the commons is thought as a verb, then its emancipatory potential is further unleashed. By establishing community-based protocols that articulate access and use, but also taking a careful and thoughtful approach to resources and distributing them in a way that focuses on the most in need first, then these acts of 'commoning' become a way of engendering the imbued prosperity of earth's resources for all. In short, a planetary commons needs to continually be 'alive' and look to move with the needs of the people and community it is serving, all the while bringing more people in. As soon as the commons becomes static, rigid and steeped in institutional wrangling, it runs the risk of falling back into capitalistic modes of operating.

So adding the term *planetary* to the *commons* forces

an expanded ontology, one that takes seriously the materiality of the commons and our inextricability from it, as well as its *ethical* potential. Elias and Moraru denote the 'ethical' nature of planetarity as infectious; a contagious way of being-in-the-world. Put bluntly, political activism is inherent in the commoning practice. The more strongly activism attempts to bring additional people and things into its ideological orbit, the more ethical potential it has. Configuring the commons as planetary, then, demands a focus on its infectious, contagious and activist characteristics and highlights how it can spread to those realms of social life that have hitherto been ravaged by capitalism.

This book therefore attempts to tease out the kinds of ethics that can aid in the flourishing of a planetary commons. It does so by offering a suite of carefully identified ethics that has the potential to articulate what a flourishing of a planetary commons may look like, what kind of characteristics it may enliven. And so the next question to ask is 'what does it mean to be ethical?'

Ethics

As Elias and Moraru have intimated, being ethical is part of the conceptualization of planetarity. It is the infectiousness of the commoning procedure. This is an important starting point for *thinking* ethically, but how can we *be* ethical? The word 'ethics' is used in many fields. There are medical ethics, which operate to guide physicians and other health professionals in their work. All new doctors are required to take the Hippocratic oath, stating that they will do no harm and putting patients' needs above any other consideration (with or without personal protective equipment). There are legal ethics, which are a set of codified rules, often enshrined in particular national legal frameworks and enforced by dedicated institutions. Within the university context, there are research ethics that each project has to adhere to. Students are given ethics forms to fill out when they are proposing their dissertations. There are business ethics that often appear as 'codes of conduct' that a corporation may (or may not) choose to enforce on its employees. People even talk about corporate ethics, which are a tacit 'agreement' of sorts that companies and costumers will enter into when transacting. None of these are the ethics that this book will focus on. In fact, these 'ethics' are the antithesis of the kind of ethics that will be explored.

This book builds on the idea of ethical thinking as infectiousness, but colours this with the conceptualization of ethics as immanent, and not beholden to any predefined higher power. Essentially, ethics are *allowing for the continuation of commoning*. Hence, they are not a 'code of conduct', but a suite of positionalities that catalyse a planetary commons wherever and however they unfold in real time. If commoning is the realization that an oppressive ideology can be resisted, then ethics are 'soft articulations' of how to maintain this resistance. Ethics therefore are 'immanent' and always unfolding, rather than some suite of transcendental ideals that are predefined. They take stock of a situation, and are subsequently articulated depending on how the commoning is unfolding. To be 'ethical' in this sense is to be guided by the given situation in all its diversity, density and difference, rather than any preconceived or external 'guide'. Such ethics are fleeting in action, but can become more durable and embedded in the world as they diffuse through the social fabric. This reading of ethics is in the tradition articulated most forcefully by the French radical philosopher Gilles Deleuze, in his sole-authored work but also in the *magnum opus* he wrote with his co-conspirator Félix Guattari.[26] In intricately analysing the notion of ethics within this work, the feminist philosopher Tasmin Loraine has argued that 'Deleuze and Guattari's conceptualisation of an immanent ethics calls on us to attend to the situations of our lives in all their textured specificity and to open ourselves up to responses that go beyond a repertoire of comfortably familiar, automatic reactions and instead access creative solutions to what are unique problems.'[27]

The 'problem' of capitalism is far from unique – it is global in its imposition – yet an ethics that aims to resist the myriad of injustices and inequalities can follow the same immanence outlined by Deleuze and Guattari. In other words, opening up spaces to allow recognition and indeed a celebration of the different forms of living justly in this world, beyond the totalizing hegemonic force of capitalism, is an *ethical* act.

As discussed above, thinking the commons as planetary entails thinking them ethically, in that they are always immanent and unfolding; they are always in a state of becoming the commons in conjunction with a community; the act of commoning. Indeed, Gibson-Graham, with their articulation of commoning, invoke a Deleuzo-Guattarian reading of becoming, arguing that it aids in producing a 'generative ontological centripetal force working against the pull of essence or identity'.[28]

Therefore, I want to build on this lineage of thinking ethically that refuses totalizing predetermined forces that can restrain difference, and instead embrace the variance that exists in the world in any given situation, in any given time, in any given place. The ethics I want to outline, then, are first and foremost grounded in this ontology; they are 'soft' articulations. They are articulations in that they can be communicated (i.e. via this book) and they are soft because they are malleable, porous and transmutable (i.e. they are necessarily 'open-source' – they can be used, adapted and transmogrified). They are behavioural and emotional 'patterns' that are constantly in flux, rather than rigid templates to adopt.

What is important to factor into this discussion, however, is that the actualization of this reading of ethics is always related to an *event*. The ethics are not predetermined or imposed; an event always happens first. In other words, for ethics to have a grounding or indeed something to be ethical towards (other than complete nihilism), they are tethered to, and unfold from, a pre-existing 'event'. As Deleuze has argued, 'ethics is concerned with the event; it consists of willing the event as such, that is, of willing that which occurs insofar as it does occur'.[29] But just want do we mean when we say 'an event'?

The Covid event

An event is when something happens that is so extraordinary that it changes the entire way everything – society, politics, economics human and nonhuman behaviour – is. More than that, though, an event is *creative*. It brings into existence entire ways of being in the world that simply did not exist beforehand. Some of these exist only as possibilities, or possible possibilities. An event emerges unexpectedly as the 'old' world ruptures, bringing new subjects, new truths and radically different experiences into existence, and shifts how that world works in its entirety. Deleuze would argue that events are 'eruptions' within a collective that calls for its complete transformation; in his words, they 'overthrow worlds'.[30]

Introduction

But such ruptures happen often. What makes an event an event is what happens after. How people react to the event maintains its 'eventfulness'. The radical change that an event brings upon the world happens through the actions of the people, communities, institutions and things that are compelled to advocate and affirm the newness that the rupture has exposed to the world. It becomes a cause to fight for, something to believe in, a truth that must be defended.

There are many revolutionary episodes throughout modern history that have been revered as exemplar events: the Paris Commune, the Russian Revolution, the Chinese Cultural Revolution, the 1968 uprisings in Paris and, more recently, the Arab Spring.[31] Events are indeed revolutionary (rather than evolutionary) precisely because they change the whole makeup – governance, behaviour, attitudes and politics – of society. The 'new' things that an event creates, then, are new voices for those whose voices have been silenced, hope for those whose hope has been oppressed, and opportunities for those made destitute. Events therefore are radical acts that bring new forms of justice into what is an unjust world.

Yet there are some 'ruptures' in the world that may have equally devastating effects, but are not events. For example, while 9/11 was a catastrophic, unexpected event (at least outside of the small group of terrorists who perpetrated it) and instilled shock and terror on a whole new scale, it served only to shore up existing geopolitical injustices; indeed it catalysed them. It foregrounded a renewed military expansion in the Middle East, and justified invasions that were more about securing oil production than bringing down dictatorial regimes. It ossified American imperialism in both geopolitical and economic terms. Crucially, then, an event cannot be known in advance, and it is only thought of as such retrospectively. And make no mistake; we are *potentially* living through what we may eventually conceptualize as an event.

In the wake of the coronavirus pandemic of 2020, the relatively smooth functioning of contemporary capitalism has severely ruptured. In the process, many different 'realities' are coming to the fore. First and foremost, we have seen the increased prominence of distinctly state-led programmes of emergency response that have gone against the grain of the capitalist policies that characterized governance structures pre-pandemic. Socialist-leaning policies have been imposed, such as universal basic income (which has been implemented to varying degrees in Spain, Germany and the UK), the rapid acceleration of the greening of our urban spaces, the championing of nationalized infrastructure programmes such as broadband internet for all, and, most importantly, the massive swelling and pedestalling of nationalized healthcare. Debt has been cancelled, and reparations to the global south have been forwarded. Also, we have witnessed the swift rebuttal of the unjust characteristics of corporate capitalism; for example, the clamour for the super-rich not only to pay their fair share of their staff's wages as they all had to go on furlough, but also to pay more tax in general. (This has been a long-standing demand, and has gained substantial traction in mainstream narratives and political discourses.)[32] The failure of just-in-time production to cater for emergency food provision in supermarkets across the UK led to food shortages. These shortcomings were then redressed by the explosion of mutual aid programmes, which also occurred across Europe and the US.[33] Street homelessness in the UK was reduced to nearly zero in the space of two weeks as those on the streets were hurriedly given shelter in empty hotels and hostels (this after years of different govern-

Introduction

ments trying and failing to reduce homelessness via various different housing policies). This surge of progressive and common politics evoked communist and anarchist tendencies that are bringing to life revolutionary paradigms of societal organization that are being championed as real alternatives to the capitalist system. As the novelist Arundhati Roy wrote (in the *Financial Times* of all places), the pandemic is a

> portal, a gateway between one world and the next. We can choose to walk through it, dragging the carcasses of our prejudice and hatred, our avarice, our data banks and dead ideas, our dead rivers and smoky skies behind us. Or we can walk through lightly, with little luggage, ready to imagine another world.[34]

Hence the pandemic could well be an 'event' because everything about the current status of the world is rupturing; the horizons of infinite possibilities are opening up. The virus supposedly came from capital's continual intrusion into the 'natural world',[35] and cares not if the victims are billionaires, homeless, the president or a prisoner. Viruses are neither dead nor alive; human nor nonhuman. A global pandemic has been foretold for many years, but nothing like this, with its global reach and rupturing force on the contemporary socio-economic fabric, has been felt before. As such, there has been an outpouring of empathetic responses, mutual aid networks, community action, the denouncement of anti-immigrant sentiments and a broader questioning of what was previously seen as the immutable status quo of capitalist realism; indeed, the activist Rebecca Solnit wrote that 'the impossible has already happened'[36] and that for the first time in a generation, we can begin to hope for a life beyond the injustices of capitalism.

Events therefore open up, rather than shut down, voices

that have been marginalized by the hitherto prevailing order of life – capitalism. More importantly, all these empathetic responses have been examples of commoning in action. It is people – from the margins *and* the centre of capitalist society – sharing resources, free from the lure of capital and markets, to meet the specific needs of the vulnerable, and these responses are creating a commonwealth that is in direct contrast to the privatized world of capitalism. They are attempts to level up the injustices of the capitalist world. The outpouring of community care, mutual aid and solidarity in response to the coronavirus pandemic came about precisely because capitalist structures were so ill-equipped to do so. The Covid-19 rupture in the human and the nonhuman world is an opening up of a portal to the *potential* of a more just, equal and common world.

However, events are also events because everything new is realized, including unspeakable things. As such, this rupture in the capitalist Pandora's box has also released untold horrors onto the world. Far right populism, peddled by a techno-fascism, is morphing into state authoritarianism and taking hold in previous Western bastions of (neo)liberal and parliamentary democracy. Even before the pandemic, horrific narratives that were considered unspeakable decades ago are now almost mainstream again, with openly fascist, racist, eugenicist and genocide-evoking rhetoric creeping back into view via social media, riot-inducing presidents and click-baiting news outlets. Throw into that potential climate catastrophe and ever more-sophisticated artificial intelligence that threatens to outstrip human ingenuity, and there is violent turbulence in the world.

In response to the coronavirus, some right-wing commentators have questioned whether taking such a massive economic hit is worth the lives of a few thousand old people.[37] The then-US president Donald Trump repeatedly

Introduction

attempted to argue, against scientific evidence, that people should 'go back to work', because the cost of the prevention of coronavirus (i.e. a massive economic recession due to the lockdown) was worse than the cure. Full-on authoritarian dictatorships, such as Orbán's Hungary, have risen up in the bastion of neoliberal democracy, the EU. In Israel, the courts were shut down.[38] Police forces in the UK and US have been accused of using new government legislation to over-enforce and release their more oppressive tendencies onto the population. National policies of increased surveillance and the subsequent further erosion of civil liberties under the guise of 'contact tracing' have been implemented in China, South Korea, Taiwan and Israel.[39]

But these lurches to the right are to be expected. When the prevailing order of capitalism ruptures, those in control and who benefit from its smooth functioning will do all they can to attempt to re-establish the status quo, and deny the new emancipatory realities from becoming pervasive. Put bluntly, those in power who benefit from capitalism will not want to see more equality, and therefore fight to maintain the standard of living they have become accustomed to.

And this is where ethics come back in. For ethics do not presuppose an externalized or marginalized 'other' to be somehow 'reclaimed'; it is not a case of positioning one form of society over another. In rebuttal of those who attempt to re-establish a totalizing narrative of capitalist realism, in remaining faithful to the emancipatory truths of radical equality that the pandemic has unleashed, we are being ethical. Simply put, ethics are mindsets and ways of thinking, behaving and acting within society that help us to resist those who look to maintain the totalizing metanarrative, and in turn help us maintain fidelity to the truth released by the pandemic event.

Where does this situate the commons? If ethics thought of in this way rebuke any kind of totalizing force, then surely the commons is just another universality to be resisted? This may well be the case if the commons becomes an extension of the capitalist state system via annexation, co-option, enclosure or adaption; or if it is a totalizing view of the world. Critically, then, any expression of the commons as a superior form of organizing society embodies the very kinds of ideas that are to resisted.

And this is why the planetary articulation of the commons – with its focus on the continual co-constitutive adaption of resource and community – is ethical. It foregrounds continual and infectious exploration of more justice, more equality, more emancipatory potential. The political scientist Glen Coulthard has always maintained that any version of 'reclaiming the commons' is fraught with colonial overtones, and argues that we should 'think about the commons as a collective effort to re-establish social relationships with each other and the land that have been systematically repressed through centuries of colonization'.[40] Focusing on indigenous communities and how they are often marginalized from a 'reclaiming the commons' movement, he argues, is an affirmation of the commons as a continual practice of commoning that seeks to rupture the smooth functioning of capital, even in its most 'progressive' forms. To therefore think of the commons ethically is to articulate those commoning practices that foreground its most radically emancipatory potential. It is to keep the commons alive and immanent, and never for them to succumb to a totalizing narrative that imposes its will on others.

Hence, the version of ethics that this book propounds strives to move us away from the status quo (or 'business as usual', the mantra utilized in the wake of the corona-

Introduction

virus) of contemporary capitalism towards a more equal and just, diverse social world that is experienced and lived *in common*. The process of commoning needs this kind of ethical thinking to resist capitalism. To protect, maintain and diffuse the planetary commons that came crashing into the mainstream in the wake of the Covid-19 crisis requires *ethical* fidelities. And, I would argue, it needs seven of them. Hence there are seven ethics *against* capitalism.

Organizing towards a planetary commons

Therefore, the rest of this book offers a way in which ethics can be thought about pragmatically to organize and maintain commoning to resist the potential evil of capitalism subsuming the (natural, material, psychological, social, cultural and economic) resources that a commons can help to create, maintain, protect and diffuse throughout society.

Specifically, seven ethics of the planetary commons have been identified, which, I believe, offer ways to recognize how commoning occurs, how it can be catalysed and, crucially, when it is being co-opted, or enclosed by the prevailing order of capitalistic ideologies and succumbing to a totalizing 'evil'. These ethics have not been arrived at arbitrarily; each has been considered as part of a layered whole that together forms a broad suite of resources. Each ethic is haunted by its opposite, a mirror image ready to subsume it if it is not practised 'ethically'. One thing that they are not is a framework, or scaffolding or an instruction manual. If anything, they act as a trellis, encouraging continued commoning above, over, through and beyond (they are open-source). Each ethic builds upon the last, but they can also be read separately as they will apply variably in different situations, places and events. So through the ethics of mutualism, transmaterialism, minoritarianism, decodification, slowness, failure and love, this book aims to propel commoning as a productive force for change in the/a/your/our world.

Ethic 1: Mutualism

It is no secret that the works and philosophies of Ayn Rand have had a profound impact upon our world in the twenty-first century. Influential politicians on both sides of the Atlantic have revealed being heavily influenced by her work, and even the personified apotheosis of anti-intellectual populism Donald Trump has admitted that *The Fountainhead* is one of his favourite works of fiction.[1] Rand's work is also said to have been a big influence in the lives of many of Silicon Valley's potentates. Steve Jobs, Peter Thiel, Travis Kalanick, Elon Musk: they have all at some point in their lives expressed admiration for the work and thinking of Ayn Rand.[2]

In Rand's *Atlas Shrugged*, the hero of the story, John Galt, proclaims in a famous speech: 'By the grace of reality and the nature of life, man – every man – is an end in himself, he exists for his own sake, and the achievement of his own happiness is his highest moral purpose.' Galt's lengthy speech is essentially Rand's manifesto of her philosophy of 'objectivism'; that the reality of the world can be directly perceived, and that any morality and social form must stem from the self's own perception of this reality. In essence, it eulogizes selfishness as the driving force of a fairer society. It rejects community, religion and government as they leach power from the individual and act as checks on the true potential of man.

It is easy to see why such a philosophy is attractive to technology entrepreneurs and career politicians of the right, and not just because of the highly masculine language; objectivism essentially gives a moral pass to selfishness and sociopathic tendencies. And such a belief in the emancipatory power of the individual over the straitjacketing of the collective is a fundamental kernel of the kind of unfettered capitalism that the paragons of nationalist populism are leading us towards. Yet it is the kind of thinking that has not only catalysed our current global tumultuousness; it is wholly unqualified to counter it.

The toxic mix of an economy of technological determinism that entices us to be more introverted, competitive and singular, with government policies that remove welfare and social security and encourage the privatization of everything, has created a world in which any kind of connection with someone else has become a transaction. Even some of the most intimate encounters we have as individuals with someone else are cast as ones in which the individual benefits. In other words, how I relate to and belong to the rest of our species becomes first and foremost a question of 'what do *I* get out of this?'

To engender a world in which the injustices of contemporary capitalism are countered, then, this Randian self-interest and individualism need to be completely banished from central streams of government, business and society in favour of *mutualism*. Furthermore, such a radical mutualism can begin to foster a world in which the collective is cherished via the empathetic encounters we can have with one another. In other words, the energies that are currently directed towards rewarding the self, when

Ethic 1: Mutualism

shared, reward everyone to a far greater degree. Hence, this ethic of mutualism argues that renouncing self-interest proposes a radical connectivity with each other as *mutuals*.

A history of self-interest

To first banish individualism, though, is to know its ideological roots. Rand's work (and the broader philosophy of neoliberalism, which will be discussed latterly) is perhaps its zenith, but it has a long history that has defined the morality of our world for multiple millennia since the Neolithic Age. I would, however, like to start in the crucible of Western thought: ancient Greece.

Thucydides was an Athenian general heavily involved in the Peloponnesian War in the fifth century BCE. As the 'father of political realism', he saw individualism as the driving force of his generation's civilization. His view of Athenian democracy was one where the strong should rule the weak, as only the powerful had the necessary qualities to lead society progressively. He was also a staunch realist, believing the world to be made up of observable phenomena, strictly focused on cause and effect, and disregarded any other kinds of extra-perspective phenomena.

Thucydides' philosophical view on the importance of self-interest as the driving force of progress, democracy and civilization was a sentiment that influenced a number of key thinkers of the 'Enlightenment' period. The Enlightenment thinkers of this time were reacting to the dominance of the church and the doctrinal view of humans' submission to a divine ruler. That all of humanity was 'fallen' and salvation was only to be bestowed by the Almighty was now seen as almost heretical, and denied the innate ability of humans to forge their own futures. So it is easy to see why

Thucydides' ideology of rationalism and self-interest percolated through to the Enlightenment's key philosophers, such as Hobbes, Locke, Rousseau and Wollstonecraft, who sit in the canon of Western Enlightening thought alongside scientific theoreticians such as Galileo, Newton and Wren.

I don't wish to delve into the countless volumes of incredible scholarly analysis that occupy entire library shelves (and indeed dusty subterranean computer servers) devoted to the intricate nuances of these thinkers' theoretical minutiae. But I hope it would not be too glib of me to say that many of these Enlightenment thinkers (although by no means all) utilized the progression of scientific, philosophical and/or mathematical scholarship to question the rule of faith in their societies, and the majority championed human nature and creativity as the source of civilization, freedom and progress. Humans were rational, empirical, inquisitive, autonomous and ultimately capable of forging their own path in life. The individual was the master of his of her (but of course at that time as of now, mainly his) own fate.[3]

These philosophical ideas matured through the centuries in the vernacular of liberalism. This was articulated most famously by Adam Smith, when he published *The Wealth of Nations* in 1776. In it, Smith argued that if everyone acted with self-interest, then an 'invisible hand' would guide us all to liberty and more socially desirable ends. 'Progress' therefore could only be achieved if we all acted with self-interest. Indeed, he argued, 'by pursuing his own interest [a man] frequently promotes that of the society more effectually than when he really intends to promote it'.[4] Smith was clear that by each being self-interested for one's own security and gain, society would be 'promoted more effectually' and liberty would prevail. He dismissed overt and conscious attempts at societal provision (such

Ethic 1: Mutualism

as governments) as ineffectual and a waste of resources. The 'invisible hand' metaphor (which in fact Smith used only a handful of times in his writings) has been used most vehemently by modern-day economists as a means to describe the efficiency of market provision. Now, it has been used as shorthand for the naturalized mechanisms of the market. Subsequent neoclassical and laissez-faire economics has distilled Smith's liberal theory of self-interest as the 'naturalized' way in which supply and demand, when tuned efficiently by people acting as *homo economicus*, should form the basis of modern socio-economic provisioning. Rational logic was eclipsing spiritual faith as the naturalized order of societal progress.

The rise of international mercantilism actualized many of these economic rationalities. It proliferated the opportunities to enact these economic ideals across vast distances, and large global companies rose from the massive financial gains. The East India Company, for example, at its height accounted for half of the world's trade (something Jeff Bezos can still only dream of), and did so through the intense trade of imported goods, mainly from India. Enacting the basic principles of self-interest and competition, the company became a key player in the expansion of the British Empire. Global trade of goods from the Far East, the Americas, India and Australia fuelled the growth of markets for more 'decadent' items in Europe (including slaves), and ushered in the dawn of contemporary globalization. The Industrial Revolution allowed for the more efficient production of manufactured goods, thereby expanding the opportunities for market growth even further. Once the US enters global trade during the turn of the twentieth century, modern capitalism has spawned a near globally ubiquitous marketplace.

At the epicentre of this market expansion was the

philosophy of Thucydides' core belief that self-interest was the only way to progress as a civilization; more wealth equals more civilization. The economic systems that built up around this core belief, the processes that powered the expansion of markets and the importance of private ownership, and the belief in the 'natural' reallocation of wealth or trickle-down effects were central tenets of the progression of capitalism in the twentieth century as it spread around the globe.

However, it was not always so smooth. A series of crises in the global capitalist system through the first half of the twentieth century began to destabilize the foundational beliefs in its robustness. The Great Depression of the 1930s showed that capitalist systems could fail spectacularly, resulting in gargantuan levels of national debt in the Western world, mass unemployment, deprivation and starvation. The subsequent rise of fascism in Europe and World War II were seen by the Smithian devotees as outcomes of the decline of civilization's belief in the centrality of self-interest and free-market capitalism as the guiding hand of progress.

The onset of political 'meddling' by sovereign states (or the too-hasty implementation of Keynesian economic policies after the Great Depression and post-World War II) was characterized by bureaucratization, state management structures, nationalized investment in infrastructures, the welfare state and notions of the social that began to encroach upon the perceived smooth functioning of the economics of self-interest. Such processes were inefficient because they were anti-competitive. If Smithian economics taught us anything, it was that unregulated and unfettered competition was the engine of societal progress. If the markets failed to provide for all, it wasn't because they included systematic inequalities; instead, failure of the

Ethic 1: Mutualism

markets was purely down to their manipulation towards a more socialized or 'common' goal. In other words, people simply weren't being selfish enough.

So while self-interested individualism propelled economic growth and created vast surplus wealth for the globally mobile capitalist class, it had not infiltrated political and social life, at least not yet. Overly bureaucratic states were not operating under the same logic, and so were hampering further growth and progress. In order to overcome this, the logic of competitive self-interest needed to be extended further, beyond economics and into the heart of government and society itself.

Neoliberalism

At least, this was the view of economist Friedrich Hayek; in the 1920s he was a prominent member of the Chicago School, a foundational body of scholarship of neoclassical economics. Hayek, a key protagonist of many of the Chicago School's main outputs, saw the creeping socialization of markets as something to be resisted. The social and political (i.e. the collective consciousness) was to be kept separate from the economic (self-interested). The conflation of the two was what had produced the ills that plagued Europe and the US at the time. Hayek, along with many of his peers (including another important figure, Milton Friedman) and students, were adamant that the autonomy of markets driven by total self-interest was a revival of the liberal project articulated by Adam Smith. Hayek's insistence upon self-interest and individualism as the path to freedom, as highlighted in his most popular work, *The Road to Serfdom*, in 1944, is a fundamental element of what we think of as neoliberalism today. Competition, as Hayek was at pains to point out, 'is the only method by which our activities can be adjusted to each other without coercive or arbitrary intervention of authority'.[5] Everyone who enters into the competitive negotiation is given a chance to decide on whether the resultant gain is worth the risk connected with it. Competition is the mechanism by which self-interested motivation can forge a path to freedom, to liberty. It was how newness and progress were to be achieved; for Hayek, 'competition [w]as a discovery procedure'.[6]

This philosophical viewpoint was a revival of the liberal project, but it would be a stretch to say that this was neoliberalism as we understand it today; the 'neo' in neoliberalism pertains to a much broader set of ideologies that utilizes Hayek's basic assumptions, but brings them to bear upon concepts beyond economics. Political theorist Wendy Brown's intricate dissection of neoliberalism argues that it 'configures human beings exhaustively as market actors'.[7] Essentially, the neoliberalism of today seeks to bring the autonomy of market mechanisms to bear on non-market institutions and non-economistic happenings. Everything can and must be reducible to the calculative mode of allocation that the market has perfected. Every decision made in our lives – be that at work, home, leisure, emotional, even unconscious – becomes an exercise in cost–benefit analysis.

The intellectual project of defining neoliberalism is vast and ever changing, arguing that as an ideology it is place-specific, and mutates across space and time. In the UK, the application of neoliberalism is most associated with Margaret Thatcher, who became the UK prime minister in 1979. She was a keen admirer of Hayek's philosophies, and set about enacting neoliberal policies swiftly and abruptly, under what the cultural theorist Stuart Hall labelled 'Thatcherism'.[8]

Ethic 1: Mutualism

Thatcher, like Hayek, saw self-interest as critical to 'social' life. I use scare quotes for 'social' because Thatcher famously said 'there is no such thing as society', but this quote is often taken out of context. It comes from an interview Thatcher gave to *Woman's Own* magazine in October 1981. In it, she castigates people who look to the government to solve their 'own' problems. People who are homeless and then look to the government to house them, or people who get a grant to cope with the problem they have; she bemoans these individuals looking first and foremost to the government, and not themselves, for help. The famous quote (which is easily available online) is:

> There is no such thing as society. There is living tapestry of men and women and people and the beauty of that tapestry and the quality of our lives will depend upon how much each of us is prepared to take responsibility for ourselves and each of us prepared to turn round and help by our own efforts those who are unfortunate.

Her notion that there is 'no such thing as society', then, really is the belief that individuals must 'look to themselves first', before thinking of others.

Thatcher's remit was the total imposition of Hayek's neoliberalism, with its cornerstone of individualism and competition, on all aspects of social life, including the family. Rather than seeing the economic separate from political, social and cultural life (which hampers growth and profit), Thatcher wanted to export market logics of competition to all corners of life. The very idea of society, collectives or the commons had deleterious effects on the power of individualism as an ideology for progress. So she imposed market logics and languages on as many different institutions and realms as possible (hence the mass sell-offs of public institutions). If every facet of life was governed by neoliberal ideology, where would alternative modes that go against this ideology gestate? Indeed, she coined the phrase 'there is no alternative', an idiom (shortened over the years to TINA) that has been utilized as a motif for the belligerent belief that marketized, competitive self-interest is the only way forward. Such a belief can be summed up by a response she gave in an interview with the *Sunday Times* in May 1981. When asked about what she intended to do in the rest of her parliamentary term economically, she replied that economics is merely the method; 'the object is to change the heart and soul'. There can be no more apt sound bite to exemplify Thatcher's belief in the ubiquity of market ideologies and the imposition of neoliberalism on all forms of life, to the detriment of the commons.

Thatcher's 'brand' of neoliberalism was mirrored in the US by Ronald Reagan, a fellow devotee of Hayek's philosophies. Regan was sworn in as the president of the United States in January 1981, after which he set about enforcing what became known as 'Reaganomics', which essentially involved large tax cuts and the reduction of government spending. Reagan's belief in competitive self-interest was therefore coupled with a generalized mistrust of notions of 'society'. This dismissal of the social as a viable realm to be nurtured and protected formed the foundations for his economic and social policies once in power. Like Thatcherism, Reaganomics is a suite of economic policies that puts competitive self-interest before a notion of the common society. In Thatcher's case, it was through the extension of the market model into social and family life; for Reagan, it was employing trickle-down economics as the basis of resource allocation.

This brand of neoliberalism, however, was not confined to conservative political hues. Bill Clinton's Democratic

Ethic 1: Mutualism

presidency and Tony Blair's New Labour in the 1990s and early 2000s recast traditional leftist party politics as more business-friendly and amenable to neoliberal policies. Capitalizing upon the shifting trends of deindustrialization and the rise of the creative industries and knowledge economies, they fully embraced neoliberal ideologies, but added a veneer of social sympathies so as to garner the popular vote (and hence sweep into power). But despite the rhetoric, they continued to champion self-interest via the language of entrepreneurship, innovation, disruption, creativity and deregulating markets, particularly in the banking industry, ultimately leading to the financial crash of 2008. After which there were plenty of activists, politicians and community groups that saw this as a potential turning of the tide on the rampant neoliberal capitalism that was engulfing the world.

But rather than this sounding the death knell of neoliberalism, its ideologues were able to point to the failure of their political opponents and blame them for the crash. All the while, however, they engaged in what the political theorist Naomi Klein called 'disaster capitalism';[9] that is, using the aftermath of the economic catastrophe to deregulate even further, decrease social welfare via austerity, and pave the way for a new populist neoliberalism that saw the rekindling of fascism all the way to the most powerful office in the world.

And it is this new, perhaps more intense neoliberalism we see today under the revivified influence of Ayn Rand. Objectivism clearly aligns with neoliberal thought and Hayek's fundamental beliefs. But Rand would go a step further when it comes to the 'interference' of government and any collective or organized forms of morality. Indeed, in 1981, one year before she died, she denounced Reagan's presidency. She took aim at his aligning with what she called the 'moral majority', namely the religious right that made up his core base; she decried it as an unconstitutional union that was a threat to capitalism. The use of anything other than the individual as a steer of progress would be mired in 'God, family and the traditional swamp'.[10]

In the aftermath of the financial crash, with the crisis of the globalized, Clinton–Blair style of neoliberalism giving rise to populism in the US and Europe, Rand's ideas are finding new traction. So as we career into the third decade of the twenty-first century, the neoliberal project that started nearly half a century ago is far from being dead, as many commentators would claim.[11] It has simply mutated. Forming an alliance with the ideology that is most readily concerned with its survival (namely the populist authoritarian right), it is still fundamentally concerned with the individualization of life, because socialized mutualism, with its focus on others, is the antithesis of self-interested profiteering.

So, beginning with Thucydides, there has been a long line of ideologues that, from a position of power, have massively altered societal systems in favour of promoting self-interest. The logic of the market, competitive individualism, entrepreneurialism and neoliberalism has pervaded national politics across the world. The fall of the Berlin Wall in 1989 and the demise of the USSR were for many, not least Reagan, the final nail in the coffin of a perceived communist 'alternative' to capitalism's global dominance. Today, neoliberal capitalism with its embedded market logics, individualism and self-interest is almost totally ubiquitous – so much so that it is an accepted, natural truth of life. Capitalism has become completely naturalized. It is an infinite and boundless domain to which there is no alternative.

Ethic 1: Mutualism

Mutualism

Yet, for all the pontificating by philosophers, for all the posturing of politicians, for all the crowing of corporations, there is very much an alternative. Realizing this alternative is how we can begin to think of the planetary commons. Questioning the pervasiveness of individualism and self-interest as the only path of societal progress and enlightenment goes against all the political, economic and cultural narratives that we see and hear on a daily basis.

In respect to how mutualism can be considered an alterative ideology to today's neoliberal capitalism, the work of the anarchist philosopher Pyotr Kropotkin is particularly useful. He was clear there was a common humanity that we can recognize at an unconscious level. And this can, indeed should, propel us to be selfless, thereby fuelling equality. This universal human connection was a force (much in the same vein as Heraclitus' theory) that can drive human progress, and one that could overcome the individualism that was rife in the political and societal theories of his time.

Kropotkin's ideas are in clear opposition to those at the core of the neoliberal capitalism of the twenty-first century. He wrote about how mutual aid – the conscious decision of a collective to work together for the benefit of everyone – among the animal kingdom and later in humans was a driving force in not only survival, but also prosperity. He saw mutual aid among animals as a principal factor in evolution, rather than necessarily the traditional view of 'survival of the fittest'.[12] Mutual aid, and its rejection of self-interest as the 'force' of evolution and civilization, is therefore an important ideology in the ethic of mutualism because it allows us to consider the rights of others in the same breath as our own, not just from sentiment, but actually as a viable means of progression. The Other is someone that can help you and augment your life via a mutual exchange of action, ideas or energies, not someone to compete against.

However, with the continued dominance of self-interested governance systems in the twenty-first century and the complete naturalization of market logics via neoliberal thought, not to mention the failure of communism (which Kropotkin's ideas related to in part), mutual aid has been relegated as an outdated, unrealistic, utopian idyll that could not form the basis of societal progress.[13] So the very first step in realizing the planetary commons is to actively shun the self-interest that Thucydides first opined on. In fact, we need to discern what we mean exactly by 'we'. Just who is included?

To delineate between 'us' and 'them', with righteousness, morality and progress on one side, and violence, greed and self-interest on the other, is to run counter to the ethic of mutualism. To think of a common world with a common humanity is to recast these divides along ideological lines rather than personal. The ethic of mutualism is exactly that: to recast individualism as part of a collective of humanity. For example, Donald Trump, for all his alleged lying, violence, ignorance, hypocrisy, insults, racism, sexism, ableism and classism, shares a common humanity, is part of the Heraclitian logos. What an ethical commoning process entails is to ensure that the ideology he espouses is not elevated to one of the highest offices in the world. What is needed is to recalibrate the dominant narrative so his views are not even given a platform, let alone an Oval Office. Indeed, commoning would question the need for an Oval Office at all. Having a social system that is based on rampant self-interest has created a 'top' and

Ethic 1: Mutualism

allowed Trump (and many people like him) to rise to it. A common world changes all that. It wouldn't even have a 'top' in the first place. Through a democratic collective will and a desire to maintain mutualism, such a hateful ideology would never be given any oxygen with which to thrive, and so ideas like that vociferated by Trump would fade away. The institutions that were built in common – media, legal, educational, governmental – would never, indeed by the very common fabric could never, allow hateful, violent or genocidal ideologies to exist. This is the task of the 'we', and it requires the efforts of *everyone*.

The science of we

John Donne wrote in 1624 that 'no man is an island'. Nearly 400 years later, it is becoming apparent, through some in-depth scientific studies, just how prophetic these words are. Because the philosophies of mutual aid, despite being antithetical to contemporary narratives of development, are finding new favour within the neurobiological and physiological sciences. In particular, there have been advances that are beginning to counter self-interest, individualism and competition as naturalized laws and realize a humanity that we all have in common; a materially existing 'we'.

For example, the discovery of 'mirror neurons' in the brain has radically altered the way psychologists view Darwinian evolution. Mirror neurons are those parts of our brain that respond to the actions of others, and then fire again when we mimic or imitate those actions. So when we see someone bang their head, we instantly feel their pain (so long as we have experienced this ourselves in the past). Or once you've experienced playing tennis, when you watch a game, your mirror neurons recreate the actions of playing tennis in the brain. The discovery of these neurons has far-reaching ramifications, the most salient of which is that they call into question the Randian premise that we are all selfish beings. In effect, the actions of mirror neurons solve what is known as the 'problem of other minds', which asks: how does one 'share' others' mental state, and make shared empathetic relations possible? The answer, it seems, lies with the presence of mirror neurons in the brain which hard-wire in us the ability to 'feel' how other people react to stimuli.[14] As such, rather than being *homo economicus*, could it be that we are in fact *homo empatheticus*?

The social theorist Jeremy Rifkin has discussed these mirror neurons at length, arguing that they are proof that that our desire to be social animals is part of our neurobiology. He argues vehemently that we are predisposed for empathy, and self-interest is a *secondary* force.[15] Being empathetic is a critical part of the ethic of mutualism, as it is the direct emotional connection between individuals that allows for shared experiences. The presence of mirror neurons helps us to do this. The connection between humans that are part of the same 'imagined community' (be that a nation, a religion, a political party, a subculture and so on) is a tangible one, but it can be coerced, manipulated and stage-managed by the state.[16]

An ethic of mutualism requires us to question this stage-managed collectivism, one that disappears at the level of empathy; it requires us to *feel* the same as others rather than simply be collectivized by a hegemonic force. This is very different from being sympathetic, which is the recognition of someone else's feelings but no direct experience of them. It is easy to be sympathetic; politicians always express 'sympathy'. But the ethic of mutualism requires more than

Ethic 1: Mutualism

recognition of someone else's disposition; it requires the experience of it. Mirror neurons help us to do this; witnessing the actions and/or plight of others fires the same feelings in the viewer. So to be more empathetic requires more experience of this. Rifkin argues that as our civilization is now globally connected, so too are our empathetic responses. So while the global hyper-connectivity of the world is being used to spread the ideology of competitive self-interest and neoliberal capitalism far and wide, it is also allowing our empathetic sensibilities to be planetary. Therefore, from the perspective of mutualism, we must continue to reject the globalization rhetoric of expansive neoliberalism, and instead be empathetic at the planetary scale.

There are countless examples of this kind of empathetic 'response' being played out across the world at the moment in the wake of the coronavirus pandemic, but I would like to focus on one from before the Covid-ravaged world. In the mid-2010s, Europe experienced the onset of a chronic refugee crisis, which saw millions of people flee geopolitical conflicts in the Middle East and North Africa, and while this has dissipated from the mainstream news agenda (for obvious reasons), it continues to this day. Families make perilous journeys across the Mediterranean Sea, and if they survive, look for a new home in European countries, many of which do not want them. Throughout 2014 and 2015, images of the refugees beamed across Europe and the world. Many of the mainstream media outlets, particularly those in the UK, maintained a narrative of national self-interest. These refugees were depicted as economic migrants and racialized Others. They were dehumanized by political and media narratives and viewed as people to resist. Couched within the neoliberal discourse of austerity, the motivations of these 'migrants' were economized in that they would be a further burden upon the already stretched national resources.

However, on 2 September 2015, an image was published across the mass media outlets of Alan Kurdi, a Syrian toddler, lying dead, face down on a beach in Turkey. The image was a shocking reminder of the plight of these refugees, and generated a distinct shift in the media narrative. Research projects conducted on social media discussions of the refugee crisis before and after this date found there to be a marked alteration in mood, from largely indifference beforehand to large-scale empathy after.[17] The image of a dead child is clearly shocking, one that unequivocally induces emotions of grief, sickness and desperation. Upon viewing it, no doubt our mirror neurons fire up, causing us to intimately feel immense grief, pain and anguish. It was a key moment (at least in Europe, the UK specifically) that highlighted how global hyper-connectivity (in this case via social media) can create an *empathetic* response that mobilizes subsequent action. Because after that date, the number of volunteers who went to help in Turkey, Greece and other areas where refugees were landing (or where they were marooned in between states) sky-rocketed. These people were (largely) spurred on by an empathetic response, a collectivized, humanitarian reaction. And in so being, they created a space of the commons. It was a commons that was both a physical place (such as in Lesvos in Greece and Calais in northern France) and a virtual space (there is a myriad of social media groups that share information and resources across Europe). While being subjected to violence by the forces of militarized capitalism and times of internal strife, a space of the commons – with people of many religions, nationalities, classes, ethnicities and genders – came to the fore.[18] The way in which thousands of volunteers and refugees can operate

Ethic 1: Mutualism

as a loose collective of people, over a long period of time, maintaining a provision of aid that feeds, clothes, shelters and serves many more thousands of vulnerable people, is an act of commoning. Such action is mutual and largely empathetic; it internalizes the feeling of the Other in a common experience.

For coming together as a form of empathetic response is a critical part of mutualism and the planetary commons it hopes to create. Indeed, scientific evidence is now suggesting that the human body can no longer be considered 'individual' at all. It has been a widely held view in microbiological science that the human body is home to at least ten times more bacteria than there are cells in the human body.[19] However, in recent years, large scientific studies have been initiated to research the so-called 'microbiome', which is 'the ecological community of commensal, symbiotic, and pathogenic microorganisms that literally share our body space'.[20] More than simply living inside us, the microbes act as another organ of the human body that is as important to our day-to-day functioning as our heart, liver, spleen or brain. Purely by cell count, we are in fact only 10 per cent human. This alien community of 'microbiota' can influence the kinds of foods we crave, the way we communicate and even the way we think and feel. It has been discovered that some microbiota release toxins that alter the central nervous system via a chemical imbalance, specifically to make us crave (and subsequently eat) different kinds of food.[21] The human body has 21,000 genes or so that all go towards making us 'who we are', but our microbiomes contain another 8 million genes, which interact with our bodies, tweak our immune system, shift the workings of our guts and brain neurons and generally help us function.[22]

The presence of other microorganisms in our bodies is

one thing; having an entire human body's genetic material is quite another. Chimerism (the presence of one person's cells in another) is a common occurrence via blood transfusions and organ transplants, as well as when, in the womb, a mother's immune system can attack that of her child or genetic material can be swapped. Perhaps more radically, it is not uncommon for the genetic code of an unborn twin to be present in the surviving child. Where one of dizygotic (i.e. non-identical) twins has not fully developed beyond the embryonic stage, their genes can cross over into the other child, imprinting their external code in its entirety. In some extreme cases, it has been know that an entire parasitic twin has been found in the abdomens of fully grown adults.[23]

All this (bio)medical evidence and research is leading many scientists to suggest that humans are part of a 'super-organism', which can no longer be delineated and bracketed off from other kinds of organisms so readily. As has been suggested:

> Whereas our cohabitation with one or another of [other organisms] may not pose a strong challenge to the commonly shared assumption that humans are unitary individuals, the presence of a large number and wide variety of such entities – and the power they have on us – renders this assumption untenable. We are not organisms but super-organisms. . . . *It is time to change the very concept we have of ourselves and to realize that one human individual is neither just human nor just one individual.*[24]

The science of mutualism may not be a 'strong challenge' to the engrained (bio)medical corpus of knowledge at present, and it may only be focused on the micro-physical connectivity that renders humanity a 'super-organism'. But couple this with leanings towards global empathy and we are beginning to see the tangible realization of a

Ethic 1: Mutualism

global, connected and collective sense of humanity. We can think of ourselves as one global (human and nonhuman) body.

Philosophical connections

The metaphor is not as prosaic as it sounds. To think of humanity as a collective, global body radically alters how we relate to one other. To shift the way we relate to other human beings – as people who are intimately connected with each other, physically and empathetically – is to renounce self-interest as the guiding force of society. But the 'traditional' view of the human body is one that implies a hierarchy. A human body usually features a brain, heart, lungs, skin, microbiota and other organs that *organ*ize that body. But what if we rethought the body *without* organs?

This is what Gilles Deleuze and Félix Guattari have done with their concept of the 'body without organs' (BwO). Put simply, the BwO is an abstract description of a more equitable, radically connected and progressive society. It is one that is a decentred, non-individualized collective and works completely differently from the prevailing order of life (in our case, neoliberal capitalism). The BwO necessitates a *radical* connectivity, one that actively shuns organization from outside. A BwO therefore has no permanent *organi*zation. Organs are created as and when needed, but it is not an *organ*ism of any kind. It is always combining, decombining and recombining to effect ongoing change to the status quo. It shares a philosophical language with commoning, precisely because it is how mutualism can work, via relentless connectivity, to create different modes of societal being beyond self-interested competition.

But thinking in the abstract in this way is not always

helpful. So let's continue with the example of the volunteers helping in the refugee crisis post-2015. It is clear that they are acting as a BwO. The response by volunteers to help refugees in Greece (but also Calais, Paris and various places along the 'migrant' route in Europe) was a long process of empathetic mutualism that spanned time and space. People from differing nations and regions would organize donations of food, clothes, tents and blankets, deliveries, healthcare and other social services for the refugees stationed in various camps across the continent. The entire operation is a BwO; 'organs' appear where and when needed (i.e. collection points, food delivery systems etc.), but it is not an organism with long-lasting centralized control (in that the vast majority of work was done without help from 'official' institutions such as governments, businesses or charities). What is more, this BwO has effected far more positive change in the various camps over Europe than many of the established 'official' NGOs (nongovernmental organizations) in the field. In this case, the body without organs was far more effective than the organized bodies.

Such radical connectivity is vitally important in maintaining this ethic of commoning. This is because self-centred competitiveness under the guise of neoliberal capitalism is always there in the background, waiting for the chance to label someone as a heroic individual, putting others before themselves. Mutualism as an ethic of the commons can result in deleterious egoism if those radical connections begin to wane. The conditions of capitalism are such that the rewards for heroic individualism increase as anti-capitalism progresses. Put crudely, people can sell out. Once-subversive subcultures that are political, anti-capitalist or anti-hegemonic are littered with examples of individuals selling out. It wouldn't be in the spirit of

Ethic 1: Mutualism

this book to rattle off names, but suffice to say there are countless examples of individuals who once championed commoning and activist subjectivities, only to be lured 'back' into the system by the promise of wealth, fame and fortune. The co-option by capitalist discourse is only too quick to pounce on 'heroic' individuals who are resisting the system. Indeed, advertising companies, branding initiatives and PR agencies constantly seek out people that would make their product stand out in a crowded marketplace, and what better way to do that then to co-opt the image of resistive behaviour?

Also, being selfless all on your own can result in exhaustion and burnout. The BwO is a dangerous collective as much as a constructive one.[25] Relentlessly being empathetic, mutual and helping others without communal support that nourishes the self results in a massive depletion of emotional and physical resources. Sometimes self-care is critical (see Ethic 7 for more discussion of this). Again, there are countless examples of lone activists who eventually succumb to fatigue, mental illness and breakdown. For those refugees and volunteers working in the camps across Europe, simple things like not only the lack of sleep and constant attempts to satisfy the needs of all those who asked, but also self-centred desires (like finding a warm, comfortable bed and a hot shower), all threatened to derail the commoning actions. However, it was through *collaboration* that these threats are (mostly) overcome. Other (less tired) people stepping in to help, words and actions of encouragement (a simple hug goes a long way), the chance to rest and have lunch with friends and other refugees: there have been countless examples of how a radical connectivity across a continent brought about a revivification of mutualism. People helped each other out as much as they possibly could.

Seven Ethics Against Capitalism

To effectively engage in the act of commoning requires a radical resistance to the dominant 'civilizing' force of self-centred competition. Its latest incarnation, neoliberal capitalism, has exacerbated the conditions of self-interest by increasing the rewards for greed (by centralizing wealth) and reduced the ability of a commons to flourish. The first step in building the commons and effecting more social justice is to inculcate an ethic of mutualism that seeks to be selfless and empathetic with others. What is more, this allows us to radically connect with others to maintain this mutualism and gain the energies of others willing to help, to stave off co-option and rescue people from burnout. It is by being together, free from a marketized transaction, that we create a productive energy that can flourish in the face of co-option and appropriation by capitalism. Such energy is emergent, always in flux and elusive. But by simply being together and open to connection, we can begin this commoning process. As the famous saying goes (with, rather aptly, an unknown source), 'be careful with each other, so we can be dangerous together'.

Ethic 2: Transmaterialism

In his influential book *Neomaterialism*,[1] the curator and Marxist scholar Joshua Simon argued that the debt-riddled neoliberal capitalism of the late twentieth and early twenty-first century has commodified absolutely every*thing*. It has stripped all objects of any ownership we as individuals or as a collective have over them, and imbued them with pure capitalist relations. Because of the explosion of consumer debt from the 1970s onwards (further catalysed by the 2008 financial crash), many of the material things that we think we 'own' are in fact our owners. Mortgages, cars, things bought on credit cards: they are owned not by us, but by the financial debt economy, and they are 'at risk if we fail to keep up repayments'. At the same time, branding has overtaken the material function of objects (a pair of Nike trainers are Nike first, and trainers second). And even within the art world, contemporary art has dematerialized the very essence of creative practice; what constitutes art becomes a question of the gaze, rather than the materials used. With the preponderance of so-called 'freeports' cropping up all over the world, which house massive collections of priceless art pieces purely for the tax avoidance purposes of the super-rich, art has become completely financialized.[2] All this leads Simon to argue that capitalism has reinvented the object, the product and 'thingness' altogether, and replaced it with 'the commodity'.

The commodity has replaced any other form of 'objecthood' such as product, thing, artefact and even the state of being an object itself. For example, a smartphone can be thought of as an *object* in so far as it relates to the sentient subject who uses, observes, abhors or admires it; it can be thought of as a *product* that has been created by child labourers in the Congo and overworked-to-the-point-of-suicide Chinese workers in Foxconn; it can be thought of as a *thing* or mere vessel which mutes its immediate presence in the world and sucks in alternative digital communications, contexts and meaning from all around. It could even be an *artefact* if put in another context beyond its telecommunications function, such as the art world, museum or archive. But Simon would see it and everything else as a commodity, which encompasses all of the above. To call a smartphone a thing, a product, an artefact or an object is to 'cleanse the commodity of the chains of its birth',[3] and hence it *is* a commodity, as is, according to Simon, everything else in this world including the land, the air, practice, the cosmos, sovereignty, peace, you, me – everything.

Simon is expatiating on what he calls a *neo*materialism, the all-pervasiveness that a Marxist reading of commodity fetishism attributes to *all* the materiality of the world. In other words, the capitalist realism of the twenty-first century has cast all material on the planet and beyond as exchangeable, commercial and therefore profitable.

This is the materialism that capitalism purveys. It is the reduction of the material world – including human materiality – to a plane of consistency that creates a marketplace. Moreover, capitalism's mechanisms attempt to

Ethic 2: Transmaterialism

order materiality, to give it this single use that predicates predictability and abstract use once extracted.[4] Entire industries are based on this: coal is for burning, crops are for eating, cattle are for slaughtering and precious metals are for mining. This translates into objects that, when produced via the capitalist machinery, have a single use and, that use once served, are discarded.

Furthermore, in discussing the inadequacies of the Anthropocene as a language of transformation to a better world, the (wonderfully titled) inhuman geographer Kathryn Yusoff argues that 'An engagement with materiality is . . . a source of communion with the formation of a collective, in which materiality is a site of political struggle and solidarity, rather than a constraint or brake to the political possibilities of life.'[5] Yusoff's work here is important to think of in relation to Simon's because it actually *presupposes* the capitalist stratification of the commonality of land and its transformation into single capitalist units of consumption. She argues that the land, and the layers of what she calls 'geologic commons'[6] embedded within it, are already imbued with inequalities that have yet to be revealed. The mining of fossil fuels, for example, is the process by which the human and the inhuman are delineated. This is done not only through the designation of nonhuman life that is to be converted into material for human consumption, but also throughout history. The industry has always marshalled which human and nonhuman bodies are expendable (e.g. slave labourers, indigenous communities, oceanic life) in their conversion into consumable objects (and hence which bodies are considered 'inhuman'[7]).

So building on these Marxist traditions of the commodity, there is a scholarship of materiality that sees the capitalism of the Anthropocene as an extractive machine,[8] one that dominantly and violently dictates which nonhuman life is to be sacrificed on the altar of profitability, and which is not (yet). The conversion of sedimented dead organisms from millions of years ago into oil renders them subordinate to the needs of the consuming human; the life of the Amazon rainforest that is destroyed to make way for cattle grazing or palm oil plantations is deemed not worthy. Capitalism therefore dictates *in real time* what is human and nonhuman. It continually defines what is human – to be a consuming, economically active, labouring, surveilled, non-racialized and obedient body – and what is nonhuman – a simple resource to maintain that which it deems human.

But within an anti-capitalist world of commoning, the material world is a site of political struggle, not something to be wielded for the purposes of building capitalist societies. The struggle of materiality is real, and plays out before our very eyes, but too often this struggle is never articulated as such. Hurricane Katrina, for example, caused far more damage to the Black population of New Orleans than the white – denoting that there are different forms of 'exposure' to the commons of the material land already latent in existing societies. Sites of indigenous conflict with the state – such as the Standing Rock protests against the Dakota oil pipeline in 2016 – are political eruptions in materiality that highlight the unequal agency and political 'vibrancy'[9] of the nonhuman world.

To bring a planetary commons into view, it is vital that the capitalist narration of materiality be resisted. Not only that: there needs to be an ethical submission to the political vibrancy of material, and the deep connections that we have as humans with the world around us. Rather than an Anthropocenic *neo*materialism being thrust upon us by a capitalist class that seeks to extract and destroy ever-further

Ethic 2: Transmaterialism

reaches of the planet's materiality, a *trans*materialism can resist this. As this ethic will argue, transmaterialism involves the reconceptualization of materiality away from it as subservient to the consumption patterns that capitalism requires, and instead as an equal plane of life that exists relationally with us as humans. It is the levelling up and dissipation of the human/nonhuman dichotomy that currently puts humans atop our surrounding nonhuman material. It is to *transcend* this material dualism, to be transmaterial.[10]

Hence, transmaterialism is an ethical commitment to levelling up the human/nonhuman divide so as to seek justice for those things *and* people deemed by capitalism to be on the nonhuman side of that dualism. There are a number of actualized ways in which transmaterialism can be practised to enliven a planetary commons, but in the rest of this ethic I wish to focus on just three: veganism, the right to repair, and eco-squats.

Veganism

In late 2019, going vegan became big business. No doubt in response to the shift in consumer demand for more vegan products in the face of a climate-change-inducing meat industry, a large number of multinational food corporations started adding vegan alternatives to their menus. KFC, Burger King, McDonalds, Chipotle, Taco Bell, Subway: all the big household names introduced vegan versions of their most popular dishes. On the surface, this is a welcome change to what is an extremely natural-resource-heavy industry. The UN estimated that meat and dairy accounts for around 14.5 per cent of greenhouse gas emissions annually, the same as all transportation combined.

The fast food industry in particular also uses an estimated 10 per cent of the world's water supply.[11] So having vegan options will allow people more choice in whether or not they decide to inflict damage on the environment. It allows them to consume *ethically*.

The problem is, of course, that simply adding a vegan burger to your menu doesn't change existing production methods; it merely adds to them. Allowing more 'ethical' choice in the market does not alter the way that capitalist processes continue to violently transform the nonhuman materiality of the world into profit; it merely displaces it. Indeed, the term 'ethical consumerism' has been coined to offer a way in which corporations can rebrand their existing patterns of marketing and advertising to appeal to those with enough disposable income to choose which product they decide to buy. Fair trade coffee, electric cars, organic food, zero-waste shops: there is a myriad of ways now that (often middle-class) consumers can maintain existing consumption patterns while having planet-destroying guilt tempered in the process. Yet, as philosopher Slavoj Žižek has argued, when buying these products, consumers buy their 'opposite'. In other words, in deciding to buy a vegan burger we are not only purchasing a form of planetary ethical satisfaction but also sustaining the very practices that create the problem in the first instance. This is ethical consumerism, but consumerism is intricately linked with the production of capitalism and so any ethical aspect to it is limited to the individual. Put bluntly, it helps to assuage our own guilt; it does extremely little, and probably nothing at all, to attack the core of the problem.

The kind of veganism that we've witnessed an explosion of in the last few years is a form of 'ethical consumerism'. There is, however, a broader movement of veganism which is resolutely ethical in the Deleuzian sense. The legal

Ethic 2: Transmaterialism

scholar Gary Francione has said: 'Veganism, which results in a decreased demand for animal products, is much more than a matter of diet, lifestyle or consumer choice; it is a personal commitment to nonviolence and the abolition of exploitation.'[12] These are lofty goals, no doubt, and indeed 'the abolition of exploitation' is something that many anti-capitalist identities will affirm. Anarchists, anti-racists, feminists, prison abolitionists, disability advocates – they will all profess to be tackling the exploitation of an unjust hegemonic, racist, sexist and ableist system.[13] However, Francione goes on to suggest that veganism intersects with these ideas, claiming that it is 'related to our oppression of other humans [as well as nonhuman animals] that manifests itself as racism, sexism, heterosexism, and other forms of discrimination'.[14]

So rather than being about consumer choice – an ethical *consumerism* that serves only to broaden the choice offered within the same system – veganism is about ethical *consumption*. We all need to consume to live, but rather than consuming the same things, veganism is about altering consumption patterns so as to dismantle the systems of oppression that capitalism entails. So this comes in the forms of reducing the obscene levels of consumption altogether in the global north, and/or consuming goods that are produced via anti-capitalist institutions such as co-ops, mutual aid groups or activist networks.

Veganism as praxis, then, is about justice for all, via an explicit process of acknowledging and acting upon the exploitation of animals. Often the 'consumerist' version of veganism is accused of valuing animal welfare over and beyond the welfare of the humans that are exploited in the system that produces vegan food (and is also accused of being couched in middle-class hipster whiteness). The boom in popularity of vegan food in the West has meant the rapid increase of industrialization in the harvesting of chickpeas, avocados, cashew nuts and other mainstay foodstuffs related to middle-class vegan consumption habits. But this has meant severe exploitation of farm workers in the global south. In Latin America, for example, avocado and banana farms are often run as criminal syndicates, and workers who speak up against low pay or poor working conditions will be kidnapped, tortured or even murdered.[15] Those who adopt veganism without a conscious recognition of how continuing to engage within a capitalist system that continues to exploit (more often than not Black and brown global south) workers merely continue to maintain the hierarchy between those who are considered human and those who are not.

As a form of ethical consumption (rather than ethical consumerism), veganism makes this nonhuman/human dichotomy more visible and therefore more problematic to ignore. By drawing closer to the agency of animals, it provides awareness of when dehumanization happens for violent ends. Too often, though, animal rights activism, particularly the institutionalized forms such as PETA (People for the Ethical Treatment of Animals), fail to do this. Take, for example, the PETA commercial that was due to be aired during the Superbowl in 2020, which depicted a number of animals 'taking the knee', akin to Colin Kaepernick's now iconic protest against police brutality. Quite rightly, the commercial was banned, but PETA put it out on social media regardless. Not only was this a crass co-option of a Black, anti-racist issue, but it tacitly signalled that animal lives are worth more than Black lives.

This militant animal rights activism, then, only serves to redraw the human/nonhuman dichotomy along lines the activists see fit. As Francione argues, veganism goes

Ethic 2: Transmaterialism

beyond animal rights, and instead is a philosophy that highlights animal welfare as being intrinsically linked to that of the search for justice for humans and nonhumans alike. As a praxis of ethical consumption, veganism rallies to bridge the conceptual gap between the human and the nonhuman and highlights how the exploitation of one leads to the exploitation of the Other.

So veganism propels us to think ethically about animals as nonhumans in line with human agency. But what about the rest of the nonhuman world around us? What of those animals, plants, rocks and trees that have been converted into consumable objects? This is where the 'right to repair' movement can help.

Right to repair

Those of you familiar with DIY will no doubt be aware of the Phillips screwdriver, a four-point crosshead screwdriver that quickly and easily allows people to screw into place a whole range of objects. However (unless you live in Canada where it is more prevalent), you may be less familiar with the Robertson screwdriver, which has a square head and fits into a square-headed screw. Its wider use in Canada is because Peter Robertson was a Canadian inventor, and in 1908 pioneered the innovative design of this square-headed screw. It was revolutionary at the time as the design allowed screws to be fastened one-handed and fixed the driver into the screw, meaning the driver didn't slip out any more and cause unnecessary damage.

It revolutionized the manufacturing industry, so much so that a few years later, Henry Ford wanted to use the design to manufacture the Model T. Robertson refused, so Ford turned to someone who had a similar design, Henry

Phillips, and the rest, as they say, is history. Being the quintessential industrialist of his time, Ford was clearly attempting to patent a screwdriver design as part of a monopolistic strategy; if people cannot take your product apart, they cannot fix it themselves and so have to come back to the manufacturer to buy either more spare parts or a brand new product.

The same technique is used today. Apple have their 'pentalobe' screw design that only they have the rights to manufacture and sell. In 2019, though, after much lobbying of governments, they eventually rolled back on their monopoly and allowed more independent repair companies to become part of their network of licensed repair providers. They did all this in response to the 'right to repair' movement, which originated in the US in 2012, but has grown to a global campaign that has resulted in changes of legislation across the world. In 2021, for example, the EU plan to bring in new laws around manufacturing which will state that for household electrical items, manufacturers will need to adhere to a minimum lifespan commitment and provide spare parts for up to ten years. And in March 2019, the US senator Elizabeth Warren called for a right to repair bill to help farmers repair their agricultural equipment, supported by 'The Repair Foundation' (formerly the digital right to repair group).

This is the legislative and official version of the right to repair movement, but it is a dispersed and variegated cultural phenomenon which includes community businesses, mutual aid repair networks, hackerspaces, and a range of individuals who campaign for people to have the right to repair and modify the goods they own without loss of consumer rights. There are millions upon millions of instructional videos online that can show how to repair things from iPhones to construction cranes. The

Ethic 2: Transmaterialism

phenomenon bleeds into direct action and activism with groups that illegally repurpose outdoor urban artefacts for political means. Subvertising, for example, is the illicit activity of changing advertising posters and billboards, often parodying corporate brand aesthetics to highlight their exploitative practices.[16]

Suffice to say, the 'right to repair' is tacitly present in all those who have the desire to 'own' their goods in a more caring and *ethical* way. To revisit what Joshua Simon noted, capitalism attempts to reduce everything to the status of the commodity, and repurposes the human relationship with materiality as one not of ownership, but of financing or renting.[17] The right to repair challenges this and claims our ownership of the object and its inner workings. With personal technological objects specifically (such as smartphones and smart watches, TVs, laptops etc.) there is planned obsolescence built in. Indeed, in 2018 Apple and Samsung were fined €10m and €5m respectively in Italy for deliberately slowing down their devices with software updates. But more broadly, the production process of these technological devices is such that the cost needed to produce durable, long-term and future-proof products is far too high. Not only are the materials needed too expensive, but Moore's Law (i.e. the doubling of computer power approximately every two years) necessitates flexibility in the hardware to maintain competitiveness in the tech sector.

There is the added issue that the 'throwaway' culture that late capitalism has developed (something that will be examined in Ethic 5) is not only environmentally damaging, but disproportionately affects those in the global south and/or the urban poor in the global north. The recycling centres where the two-year-old phone goes to when a contract is upgraded can be highly dangerous places where toxic metals leak into the water supply, and the urban poor will 'scavenge' around for items to circulate within the informal economy.[18] There are obvious ethical advantages to this recycling process in so far as the phone doesn't end up in landfill or get incinerated (not all of it anyway). Yet the global trade in waste itself is gargantuan. There are over 200 million tons of waste traded between nations every year, with an estimated worth of around $300bn.[19] This figure has grown rapidly in recent years as short-life, disposable goods have become the norm.

The right to repair movement is fighting against the throwaway culture that late capitalism thrives on, and pushes for more time, education and resources to delve deeper into the materiality of the things we use everyday. It opens up the 'black box' of the commodity and breathes a fresh vitality into the once single-use, statically defined object. The political theorist Jane Bennett has argued that there is a 'thing-power' to objects that can have a 'material recalcitrance', and force us as humans to change the way we behave.[20] When we open up an object and combine its material composition with our own knowledge drawn from our cultural and social conditioning, the commodities all around us suddenly can become anything we want them to be. They can be reused, recombined, reconditioned or upcycled (to use a more craft-industry-friendly term) into something completely new. In the process, the power to change us, our environment and our relation to other things is released onto the world. Taking apart a broken television and refashioning it as a table is one example, one that, while it clearly reduces the need to throw away the TV or buy a new table, is still couched in consumerist narratives. However, turning old pillow cushions, curtains and bed sheets into much-needed face masks, healthcare uniforms and other personal protective equipment (PPE)

Ethic 2: Transmaterialism

during the recent pandemic is quite another; the right to repair suddenly becomes something that could have saved lives.[21]

The materiality of the things that we consume under capitalism is disposable, unitary and devoid of any agency in the world. With a right to repair mindset, suddenly that materiality has more complexity, more use value and more vitality. Indeed, a slogan of many environmental campaign groups from around the world (including the 52 Climate Actions, a collection of ten Europe-wide scientific research institutions) is 'refuse, reduce, reuse, repair and recycle'. Designed as a hierarchy of ethical consumption of goods, the first option should be in refusing the thing in the first place. The second is to reduce your consumption of it. If you need to use it, continue to do so for as long as you can. If it breaks, then repair it. If all else fails, recycling should be the last resort, chiefly because that good enters into the global waste industrial system with all its environmentally damaging and exploitative issues. During World War II, the general public in the UK were advised via a governmental campaign to 'make do and mend' their own clothes as apparel production had been rationed; perhaps it is time to revisit such a campaign.

With both veganism and the right to repair, the onus so far has been on the individual. They both adhere to a form of nonhuman justice that has 'the abolition of exploitation' embedded within it, but they are more suited to a lifestyle narrative, something that can easily be co-opted by an appropriative form of consumer capitalism. In other words, they can both become a form of 'ethical consumerism' (as Žižek would articulate it) unless they are combined with the mutualism detailed in the previous ethic. In other words, shunning the individualism that this co-option thrives upon and being radically connected to each other *and to things around us* brings a planetary commons into focus. As Elias and Moraru have argued (and as outlined in the introduction), the planetarity of the commons rests on the ethical application of the commoning process, that is, our acknowledgement of our epistemological inextricability from the material world around us. This means bringing into the encounter with the nonhuman world all the empathy and selflessness that we bring to the encounter with our fellow humans. The 'thing-power' that rests in the object, as Jane Bennett so vividly argues, can affect our world in radical ways, and is valuable fuel for our own emotional energies. The material world, far from being inert and consumable, can power the human condition for more empathy and radical action.

This is nowhere more evident than in the many examples of eco-living, squats and communes from all over the world.

Eco-squats

The Diggers, as mentioned in the introduction, were an important group in the conceptual foundations of the commons. They believed in the commonality between humans and the land that fed them. Drawing on his own reading of Biblical teaching (something which was more prominent because of the advent of the printing press) rather than that of the clergy, which often preached subjugation to the monarchy, Gerrard Winstanley believed that everyone was created equal, and therefore had fair use of the land that God had provided. With a relatively small band of followers, he began digging up the untouched land near Weybridge in southeast England to grow corn, peas, carrots and other produce. This was then distributed to

Ethic 2: Transmaterialism

whoever needed it. For about a year or so they were able to stave off the army and the lawmakers before they were violently evicted; not, however, before they went on to create other groups, notably the Levellers, who were more radical in their practice. Foregrounding many of the democratic ideals that were to follow centuries later, the Levellers were drawn more from army personnel, and went on to create a list of demands that have characterized, to some extent at least, what we now see as parliamentary democracy.[22]

But at the root of the Levellers' demands and the Diggers' praxis was that the land was common and should be shared with everyone, rich or poor. To rebel against the monarchy and the dogmatic teachings of the church so forcefully was a radical step at the time. But while the Leveller movement was eventually quashed by Oliver Cromwell in 1649, their ideology still enlivens anarchist and anti-capitalist movements today.

The Levellers' creation of a 'commune' that provided all that life needed is a model we see today. Even with the humble urban allotment, there is a sense of tending the land, distributing the food via local mutual aid networks (indeed, during the coronavirus lockdown, allotment-grown produce was a major source of food for those unable to get to, or deliveries from, the supermarkets), and a tendency to reuse and readapt objects before recycling them via institutional systems. As well as allotments, there are many less official (often illegal) eco-squats around the world. Some of the squats that I have been lucky enough to visit include Christiania in Copenhagen, Can Masdeu in Barcelona and Grow Heathrow outside London, but there are many others across the global north and south (albeit with differing gradations of ethical transmaterialism).

They are first and foremost squats, in that they were created by a group of people looking to inhabit a plot of land which they did not own, and use it in a more socially just and equitable way. These eco-squats embody the anarchist mindset of the squatting movement more generally. The anarchist tradition has been built upon the rejection of *any* form of societal organization – *an*archy, without any form of archy, be that monarchy, patriarchy or indeed speciarchy. Far from the mainstream political view espoused by the elite (and failing presidents), anarchism is not total chaos and the absence of order. It is the desire to organize society free from any centralized or powerful control. Providing needs and meeting wants within a community of people without giving up rights, surplus or control to an externalized power is the central tenet of anarchism; and to do so actually requires a great deal of ordering, communication, deliberation and debate. Anarchism, then, is significantly more likely to be found in discussions of how to run a community garden than it is in throwing a Molotov cocktail at riot police. As an ideology, anarchism defenestrates any need for a political elite that govern us so as to exploit us; that is perhaps why political leaders are so scared of it.

This anarchist desire for the horizontalization of society extends to nonhuman and material matter. And so within these eco-squats, there is a levelling of the human and the nonhuman via very different ways of living. Take the Grow Heathrow squat, for example. It initially started out as a campaign against building a third runway at Europe's busiest airport in 2010, but has grown into an anarchist community that provides a window onto how the ethics of a transmaterialist world can be realized. Within the squat, all electricity is generated via renewable sources. There is a wind turbine, itself built out of reused materials, and a number of solar panels. All food consumed at the site is grown there, with a strict vegan diet. The community have

Ethic 2: Transmaterialism

their own apothecary for herbal medicines, they reuse and recycle every bit of waste (including human waste, which is transformed into fertilizer within six months), and even power washing machines via bicycles linked up to the machine's drum with a rubber band. And while it may seem trivial, if you have ever had to pedal your way through a standard washing machine cycle, you'll get a *much* better understanding of how much energy is needed to simply wash clothes. It gives you a more embodied knowledge of the energy that we take out of the ground just to clean your daily clothes load in a machine. Also, Grow Heathrow operates a policy that the small number of living quarters they have there (up to twenty people) are offered to those in most need because of their rejection from mainstream society, and will often prioritize the trans and homeless community (this is an example of the site's minoritarianism, which is explained in the next Ethic).

Grow Heathrow is not alone in having this ethic of transmateriality at the core of its anarchist ethos. Revisiting Yusoff's work, it is clear to see that Grow Heathrow has an engagement with materiality that is based on a deep socialized connection with the land. But in addition, the site's transmateriality is based upon overt anarchist politics, a politics that originated in protest over a third runway (which would have seen a massive uptick in the carbon emissions of an already high-emitting airport), but now buttresses a way of life that opens up the possibilities of a sustainable life beyond capitalism.[23]

This political aspect of the ethic of transmaterialism is vital; too often post-phenomenological accounts of materiality (such as those put forward by Manuel DeLanda[24] and Graham Harman,[25] often in the vernacular of actor-network theory, object-orientated philosophies and/or assemblage theory) are critiqued for their lack of political potency.[26] And with good justification – too often these accounts level up the human and the nonhuman in the desire for pure analytical consistency. But in so doing, they ignore (or explain away as simply an extension of this nonhuman/human blending) the socio-political contexts that create the injustice and inequality in the first instance. An ethic of transmateriality, then, follows Yusoff's lineage in that politics is woven through the human engagement with the nonhuman world; to interweave them apolitically is to not tackle the constant degradation and exploitation of material that has brought us to the precipice of climate catastrophe.

This does not mean, however, that these eco-squats can be heralded and pedestalled as the politically anarchist vanguards of an ethical transmateriality. Clearly they are dynamic, debated, contested and living sites and struggle to maintain the purity of anarchist life permanently and uniformly. Grow Heathrow itself is not immune to such contestation, and to fetishize its clear emancipatory potentials would be to place too heavy a burden on it as a model of transmateriality. But conversely, these sites do offer a way to show that an ethical commitment to transmateriality can, indeed must, stretch beyond individual lifestyle choices. Being vegan and attending hackerspaces is an important part of embedding an ethic of transmaterialism into our own lives. But in order for a planetary commons to come into focus, these ethical commitments need to inform broader societal and structural changes. And so these eco-squats, based as they are on how the ethics of transmateriality can engage mutualism to work together for the common good, provide a glimpse into the 'leap' between transmateriality as a lifestyle choice (which can be easily co-opted, branded and sold as just

Ethic 2: Transmaterialism

another axiom of capitalist growth) and transmateriality as a force to reject this capitalist appropriation, and to grow a sustainable community that works and lives in conjunction with the common resources of the land.

And we can see glimpses of this even on the scale of national parliamentary politics. In 2010, Bolivia passed the Law of Mother Earth, which sees the 'natural' world as the 'collective subject of public interest' and as such having equal rights with humans within legal frameworks.[27] This had repercussions more recently when, in October 2020, the socialist party Movimiento al Socialismo was voted into government, much to the ire of the billionaire Tesla owner Elon Musk, who was allegedly behind a coup in order to control the country's lithium deposits.[28] And in New Zealand in 2017, Parliament passed a law that saw the Te Awa Tupua river obtain the same legal rights as a person, something that the local Maori tribe had been insisting upon for centuries. In 2019, Mount Taranaki also gained the same legal rights. These are instances of the levelling up of the human and the nonhuman with the laws of the state apparatus, thereby signalling that if the political will is powerfully present, then a transmateriality can easily be achieved nationally, even within the confines of parliamentary democracy.

A planetary commons thrives upon the commoning practice of constantly evolving the protocols of justice and equality that are needed to evade capitalist co-option and the violence of accumulation by dispossession. Grow Heathrow and the other eco-squats around the world, despite the 'dilution' in some sense of their ethical transmateriality, can transfer that ethical practice to other parts of the world, if the political will is there to do so – as has been shown in Bolivia and New Zealand. Eco-squats can provide the means by which we can engage in a deeper connection with the material world, and they show, often in very practical ways but also in intangible ethical ways, how this can be done.

Ethic 3: Minoritarianism

The work of the late great Mark Fisher, who died by suicide in 2012, has been hugely influential for those who theorize a world beyond capitalism. One of his most-read works, *Capitalist Realism*, released in 2008, is a beautiful polemic that evidences just how difficult it is to imagine, let alone actually build, a society that is free from the violence of capitalism. Indeed, the title of chapter 1 of the book is 'It's easier to imagine the end of the world than the end of capitalism' (a phrase originally attributed to Fredric Jameson and/or Slavoj Žižek), which has become somewhat of a memeified slogan. Pictures of shoppers clutching Versace bags while knee-deep in floodwaters in Venice; amused diners in Burger King snapping a pic through the glass wall of a Parisian gilet jaune (yellow vest) protester fighting his way through a cloud of tear gas; golfers teeing off while wild forest fires conflagrate behind them; a 'Covid-19 essentials' shop opening in a mall in Miami: these have all been used with 'It's easier to imagine the end of the world than the end of capitalism' as a caption on various social media platforms. The implication being, of course, that while climate change, global uprisings and pandemics surge around the world, consumerism continues regardless.

Seven Ethics Against Capitalism

One of the most striking analytics of Fisher's book is his use of cinema to vivify his theoretical musings. The chapter that has launched a thousand memes opens with a description of Alfonso Cuarón's 2006 film *Children of Men*. The film, itself an equally brilliant work of searing artistic critique, Fisher argues is prosaic because it shows us a world that is ending – the film depicts a society in which children are no longer born, hence humanity is heading for extinction – but one in which people also still go to work, shop, are advertised to and consume; 'internment camps and franchise coffee bars co-exist'.[1]

If you have not already done so, I do urge you to go and both watch the film and then read Fisher's book, as his analysis is far more eloquent and beautifully morose than anything I could ever do (plus, I am about to give spoilers). One of the central plot themes not only is another vehicle for thinking about capitalism's trajectory, but also helps to conceptualize the ethic of minoritarianism. The central character, Kee, is the woman who we find out halfway through the film is pregnant. Kee is a young, Black, evidently working-class migrant woman. The film then pivots into a standard heist-style movie, where the male lead, Theo, attempts to get her to a secret scientific commune called the 'Human Project', while warfare, concentration camps, violence and protest continue all around them.

There is a quite ethereal part of the final act in which Theo and Kee, after she has had her baby, are exiting a building during a bout of urban guerrilla warfare. The cries of the baby stop the soldiers dead in their tracks. As Kee cradles her baby and walks past them, all the soldiers peer at the baby in silent amazement before a mortar strike signals the start of the cacophony of war once again. It's almost as if the sight of the potential continued existence of the human species is not enough to stop people from

Ethic 3: Minoritarianism

maintaining the process of state violence. As Theo leads Kee away from an activist who is attempting to take the baby into his own care, the activist shouts, 'We need the baby, we need him!' Theo simply replies, 'It's a girl.' Theo is killed shortly after and the end of the film sees Kee and her baby rescued by the Human Project, with the distant sound of children laughing as the screen fades to black.

That the film portrays the potential saviour of humanity as a young, Black, migrant, working-class woman, I don't think is an accident. It is a statement to affirm that the world of white men is violent, destitute of hope and riddled with capitalist realism. To emancipate ourselves from this dystopian future, we (the 'we' as defined in Ethic 1) must seek to be more like the minority groups of this world, and let them lead the way to a more hopeful future for our children. The young, working-class, migrant Black woman Kee and her baby signify the many subjectivities that are marginalized and oppressed in this world by institutional sexism, racism, classism and the rest. Minoritarianism, then, is fundamentally an ethical alignment with oppressed marginal subjects, whether we are ourselves marginalized or not, in order to understand how the oppressive institutions of capitalism work, and how they affect the lives of everyone. A planetary commons (and the commoning process it entails) is energized by 'new' thinking, subjectivities and drives for equality that emanate from the margins of capitalist society. The struggle for equality has always come from the margins and from oppressed people. From Moses' emancipation of the Jews from slavery in Egypt, through the Suffragettes gaining equal voting rights for woman, to the continual struggle for racial justice in the modern world, equality is not bestowed upon people from above; it is won from below.

In the wake of the Black Lives Matter movement that

started in 2014 but gained huge global momentum in the summer of 2020, the racism inherent in capitalism's long history has been starkly laid bare, but has also forged a broader critique of capitalism's articulation of which bodies are worthy, and which are disposable.

Capitalism *de*humanizes certain people, deems them dispensable and makes them resources to be consumed and/or disposed of (à la the previous Ethic). Dehumanization is a violent process that first designates animals as disposable, and second, reduces certain gendered, raced or ethnic bodies to that 'level', and therefore proclaims they are disposable as well. When politicians describe migrants as 'vermin' or Black protestors as 'animals', they are using loaded language and a broadcast platform to perform this dual process of dehumanization. With the Black Lives Matter protests of the summer of 2020 and the subsequent far right and openly fascist rallies and counter-protests, it became painfully clear all too quickly which of those groups were more harshly policed. Images of police forces brutalizing anti-racist protesters while remaining passive in the face of far right violence speaks volumes as to whose lives are deemed worthy and whose are not.

Hence this ethic of minoritarianism plays into social theoretical discussion of the dialectic between the 'majority' and the 'minority'. What is more, those most persecuted (i.e. racialized but also sexualized, classed and disabled subjectivities) have shown everyone else how to cope (and if lucky, how to survive) under the pressures of an unjust and violently marginalizing capitalism. Studying, listening to and learning about the struggles of the oppressed (through movements such as decolonizing cultural and educational institutions, transgender struggles, disability narratives etc.) will be more critical than ever.

A planetary commons cannot claim any ethical commit-

Ethic 3: Minoritarianism

ment to equity if it does not recognize the inequality that has been created and perpetuated by centuries of capitalist development. Hence, an ethic of minoritarianism does this by first realizing how the conflicting pathways of prejudice create different forms of subjugation, then engaging in intersectional thinking, and finally, and most crucially, empathetically engaging with those who have 'other' forms of subjugation than your own. It is to each of those 'steps' of minoritarianism I will now turn.

Becoming minor(itarian)

Returning to the philosophies of Deleuze and Guattari, they devote a large proportion of their corpus to the notion of *becoming*, the constant state of emerging in the world; that is, their work is concerned with the philosophy of 'the subject' and its constant creation. They are adamant that we must always engage in becoming as it is this process that unsettles and destabilizes the status quo, the core subject, the majority. But, as they are keen to point out, the majority is not simply a numerical concept:

> When we say majority, we are referring not to a greater relative quantity but to the determination of a state or standard in relation to which larger quantities, as well as the smallest, can be said to be minoritarian: white-man, adult-male, etc. ... The majority in the universe assumes as pregiven the right and power of man. In this sense women, children, but also animals, plants and molecules, are minoritarian.[2]

The majority then, for Deleuze and Guattari, is an abstract idea to which we are all beholden but never fully become. Within capitalist dogma, that is the pure *homo economicus*, the perfect neoliberal subject that will be white, male and fully in tune with market forces. It is not a simple matter of numbers; rather the majority is the dominant force of the prevailing order. The majority *actively* creates minor subjects when we inevitably fall short of this abstract ideal. And when this failure happens, we are Othered, marginalized and then labelled as 'outside' (often to then be co-opted or violently oppressed). Those with the power to narrate this majority centralize it in an elite cohort, who have the ability and mechanisms available to them to distribute the resources that are generated by co-option of new subjectivities. This has manifested across the centuries as processes such as colonialism, where rich White European powers have plundered global south countries for raw materials and peoples, asserting their white supremacy over the globe.[3] One of the most violent examples was when King Leopold II of Belgium ordered the mass genocide of indigenous Congolese people in order to essentially 'asset strip' the Congo of its rich raw materials (notably rubber) and redirect the profits back to Belgium to build palaces and fund his extravagant lifestyle. Equally barbaric was the killing of Kenyans who were revolting after British Imperial rule for decades. There are of course many, many more examples just as horrific across the 'New World' in the Caribbean, India and Asia.

That capitalism is predominantly white, male, able-bodied and heterosexual is no accident; it stems from the conquests of history and the prevailing ideology of scientific rationalism, Thucydidian notions that the powerful must rule, and European (and later pan-Atlantic) cultural and industrial hegemony. White, male, elite-class influence continues to pervade capitalist power brokers and, like the violent colonialism of the past, manifests itself in violent prejudicial affects today. It is white men who have mastered history and hence wield institutional power.

Institutionalized racism that the Black Lives Matter

Ethic 3: Minoritarianism

movement has exposed, pervasive societal sexism (exemplified by the #EverydaySexism and #MeToo movements), transphobia, Eurocentrism (such as the Mercator world map projection), prejudice against those with disabilities: these are all symptoms of a capitalist state that holds a particular kind of identity as 'powerful' and everyone else as marginal. Indeed, such a process is necessary for capitalism to grow. To continue bloating the resources of those who already have too much, since they cannot take from each other, it has to continue to maintain a marginalized 'other' to exploit, and to blame when things go awry.

So when wages drop, unemployment rises, neighbourhoods change and diseases spread, these can be blamed not on the deficiencies of a capitalist system in protecting and providing for all, but on minority subjects. This plays out in the all-too familiar trope of blaming immigrants for job losses when austerity is to blame or there is another regional lockdown in a pandemic; or, in the US, of the government blaming the rise in terrorist incidents on refugees or asylum seekers, when it is home-grown domestic terrorism that is far more deadly.

So the *minority* therefore consists of identities and experiences that have been cast 'outside' of the majority by the powerful, either as unwanted and disposable, or as a resource to exploit (via co-option and appropriation). The ethic of minoritarianism, then, involves actively creating a commoning practice that foregrounds those minor subjects and how they are 'created' via institutionalized prejudices, and makes spaces where those prejudices are resisted and nullified. No mean feat, for sure, but a planetary commons, if it is to maintain ethical fidelity to a more equitable future, needs to be undertaken by everyone including minorities, as Deleuze and Guattari would argue. They write: 'Woman: we all have to become that, whether we are male or female. Non-white: we all have to become that, whether we are white, yellow, or black.'[4] In so doing, they are asserting that in order to subvert the prevailing order, all subjects must become minor; and critically, *continue to do so*, thereby evading the co-optive nature of capitalism.

I have written elsewhere about how the contemporary forms of creative capitalism thrive on the appropriation of 'new' subjects that can be marketed.[5] For example, recent times have seen corporations being accused of 'pinkwashing' as they plaster the pride flag all over their social media feeds, all the while upholding structures that systematically persecute queer minorities. The term has also been used to describe a deliberate strategy by the Israeli government to conceal the continuing violations of Palestinians' human rights behind an image of modernity signified by Israeli LGBTQ+ life.[6] More recently, during the Black Lives Matter movement, many corporations and institutions were critiqued for joining in with the spectacle of the movement, all the while changing nothing about their racist employment practices of Black people or (for example) their use of child labour in African countries.

This co-option of minority subjects for profitable gain is standard practice within capitalism and sees many of the political and/or anti-capitalist ethics stripped away. But to engage in minoritarianism is to recognize those pathways of subjugation, reignite the ethical commitment to the struggles of these minority peoples, and resist co-option and the deadening of the critique that this ultimately brings.

Intersectionality

So enacting minoritarianism entails engaging with subjectivities and alternative imaginations that destabilize the

Ethic 3: Minoritarianism

major *and* minor, thereby dissuading the majority from taking hold. This builds on the other ways of thinking a planetary commons ethically, by orientating our (human and nonhuman) collective agency towards realizing identities that are beyond the abstract majority, beyond capitalism.

But what about the *specifics* of these subjectivities? It is clear from the deep history of imperialism and capitalism that anything that is non-white and non-male is minor. However, to aggregate the multiple subjects beyond the major is to deny their specific identity politics. Indeed, becoming-women will intersect with becoming-Black with becoming-queer with becoming-working-class with becoming-disabled in many different ways, at many different points, at many different times. They will conflate, conflict and contract in reaction to each other. This is the core work of intersectional feminists such as Kimberlé Crenshaw and Judith Butler.

Crenshaw is credited with coining the term 'intersectionality', using the analogy of a traffic intersection. If an 'accident' occurs (e.g. a Black woman is oppressed), it may not be obvious in which flow of traffic (form of oppression – sexism or racism) the accident occurred; indeed it may have been both.[7] Essentially, the multiple forms of capitalist oppression cannot be fought independently for long. Building a planetary commons will require intersectional action that recognizes not only how the feminist struggle is undertaken, but how the actions of that struggle adapt to anti-racist struggle, and then to disability struggle, transgender struggle and so on.

In a similar vein, Judith Butler has cautioned against labelling all women as 'other' to men. Butler's seminal work has problematized, and continues to problematize, gender identity. She argues for a fluidity of gender where being-man or being-woman is not a dichotomous ontology, but a performative position that can shift, morph and change in different times and spaces. Butler's work has questioned the feminist discourse as solely about empowering 'women' as a coherent bloc. She goes on to challenge the biological dualism of sex, arguing that that too is a construction of language; it is 'something like a fiction, perhaps a fantasy, retroactively installed at a prelinguistic site to which there is no direct access'.[8] So for Butler, gender (and the sexual dualism which it is built upon) is not a set of static categories but a fluid identity that is continually performed. As such, to categorize the agency of all women together in the feminist movement is to discontinue the act of becoming, which, as we now know, can lead to appropriation by the majority.

So each movement of minority struggle challenges different entities of the capitalist majority. These struggles destabilize vastly different axioms of capitalism, and create new (perhaps conflicting but intersecting) territories of identities and subjects, however fleeting they may be. This has been evident with the Black Lives Matter campaign that started in the US after the acquittal of George Zimmerman for shooting dead African-American teen Trayvon Martin in 2012. In 2014, Michael Brown, a young Black man in Ferguson, Missouri, was killed by the police, after which multiple riots occurred in US cities. Then in 2020, the images of George Floyd apparently being murdered by the Minneapolis police sparked a global movement. Beyond the mainstream media and kneeling sportspeople, though, Black Lives Matter is a progressive movement that continues to affirm the 'otherness' within Black culture. Indeed, their website states: 'We affirm the lives of Black queer and trans folks, disabled folks, undocumented folks, folks with records, women, and all Black lives along the gender

Ethic 3: Minoritarianism

spectrum.'⁹ This further affirmation of minor subjectivities within the minor subjectivity of Blackness evidences how the movement is continuing to eschew normalization, and how it is engaging in minoritarian thinking that is intersectional; in other words, the movement (and the relating literature, art works, speeches, marches etc.) is actively pursuing an intersectional agenda which seeks to acknowledge the different forms of prejudice that exist with and through racism.

The ethic of minoritarianism is therefore the thinking beyond the abstract ideals of the major *and* minor ground. It is a critical ethic in maintaining commoning because it propels us to continually enact 'new' subjectivities that are as yet unrealized by the majority (i.e. capitalist subjectivities) *and* the minority (i.e. static 'othered' subjects that await co-option). The appropriative mechanism of capitalism means that any new 'minor' subjects will be brought under the remit of marketized logics if they stop becoming. It was easy for corporations to claim 'solidarity' with Black people, but to then go further to claim it with Black trans folk too? Not so palatable to shareholders. So to continue to destabilize the minor ground, we – as defined in Ethic 1 – must continue to engage in minoritarianism, to enact and perform alternative subjectivities that are uncapturable. The more this is done *in common*, the less power the appropriative forces of capitalism have. Having this mindset allows everyone to think things and relate to people beyond the majority, to imagine new subjects that can challenge the injustices of capitalism.

Building a planetary commons with this minoritarian ethic in mind, then, radically problematizes the perceived notion of 'reverse racism' (where white people can be oppressed by Black people) or 'reverse sexism' (where men are deliberately disadvantaged relative to women).

Within the contemporary throng of modern media, there is a plethora of voices (largely emanating from the alt-right and their 'intellectual' paragons) that contend that the drive towards gender or racial equality comes to the detriment of male and/or white identity. In brief, white men are claiming that they are the ones that are being oppressed by feminist and anti-racist movements such as #MeToo or Black Lives Matter (spawning movements such as All Lives Matter in reaction).

What these movements are designed to do is to deny the long history of minority oppression (of female, non-white, disabled, queer and/or working-class identities) that has built up over millennia of majoritarian rule. To claim that (for example) men are being discriminated against in the business world because of potential laws that would necessitate 50 per cent female membership of corporate boards is an attempt to deny becoming-minor. That men make up 85 per cent of corporate board membership globally is no accident.[10] It is because of an economic system that has been built by and for masculine identities. And this system has been built upon a long history of European mercantilism that again has male white power as its overwhelming identity. And this system was built upon the Thucydidian belief that the strong must rule the weak. And this entire system is built upon the oppression of non-white and non-male identities. To claim that the last few years or so of female empowerment are somehow an affront to the white male rule that has sedimented in our politics, economies and societies over millennia is, at best, a stretch. What it is instead is the continued attempt by the majority to maintain the othering of minor subjects on which elite wealth can be built and maintained.

When taking an intersectional approach, it is important, therefore, to acknowledge not only the structural nature of

Ethic 3: Minoritarianism

prejudices (rather than explaining them away as an individual character defect, known as the 'bad apple' excuse), but that each structure of prejudice has its own particularities and nuances (racism against Blacks is not the same as racism against Travellers, for example). As such, the ethic of minoritarianism will 'look' very different depending on the time, place, space and system of oppression it is resisting.[11] To engage in the ethical practice of minoritarianism is to be conscious of this constant ebbing and flowing between different structures of oppressive prejudice. This is why this ethic adheres very much to the notion of commoning, because it is a dynamic process where we have to be sensitive to the different languages, symbols and gestures of differing marginalized groups. In effect, minoritarianism is a constant process of becoming-minor, in the Deleuzo-Guattarian vein. We have to become, and then *stay* becoming-minor. In order to explicate this process in a bit more analytical detail, I want to turn to the minor subjectivity of disability.

Staying minor

In the UK, under the austerity programme that began under the Conservative–Liberal Democrat coalition government in 2010, people on incapacity benefit were required to undergo a 'fitness to work' assessment (something which continues to this day with subsequent Conservative governments). The assessment process (undertaken by private companies) involved taking a 'test' that assessed your ability to enter the workplace or not. It asked simple questions like 'what is 100 minus 25?' and 'can you get dressed?', but failed to account for systemic, invisible, mental or complex illnesses. In March 2013, as Linda Wotton lay in a hospital bed suffering complications from a heart and lung transplant, she was passed fit for work. Linda died nine days later. In August 2013, Mark Wood starved to death at his home, four months after having his disability and housing benefits stopped. He was passed 'fit for work' despite having complex mental health problems. In September 2013, Michael O'Sullivan, a man who suffered from depression and agoraphobia, killed himself after being passed 'fit for work', despite three different GPs' evidence to the contrary. In April 2019, Stephen Smith died after years of chronic illnesses. He was 'declared fit for work' in 2017 and hence had his housing and incapacity benefits removed. Later in 2017, following the campaign of the disability activist Gail Ward, the Department for Work and Pensions released mortality figures showing that between 2014 and 2017, 111,450 disabled people died shortly after they were passed fit for work.

There are many more cases of deaths and mortality statistics that paint a sobering view of the UK's disability benefits scheme, but the evidence is clear: the UK's disability welfare is a specific manifestation of the neoliberal agenda of normalizing bodies for capitalist productive labour. Even the name, 'fit for work', is a constant reminder to people that the governmental system of welfare and disability benefits is geared towards making employable bodies. Neoliberal capitalism requires an able body to labour, so any body that cannot contribute to the continual growth of capitalism is castigated as unnecessary, and even burdensome given that they can 'distract' able bodies from their capitalist duties (i.e. carers and minders who give up work to look after their disabled loved ones). This was laid even more starkly bare during the pandemic lockdowns when older and more vulnerable people were sacrificed to the virus on the altar of 'getting back to work'.

Ethic 3: Minoritarianism

In other words, our contemporary neoliberal capitalist societies, which champion competitiveness over compassion, market forces over cultural variation, labour over leisure and self-obsession over sociality, reward the normal able body and marginalize and discriminate against the 'disabled' body or that which cannot function efficiently as a capitalist worker. Being deaf, blind, mute, a wheelchair user, autistic, bipolar or any other of the myriad of alternative corporeal and neuro-diverse states is indeed disabling within these societal parameters, as people with these varying corporeal states will suffer prejudice, discrimination and injustice (as we see with the UK's disability benefits controversy).

To resist this process is to disarm this weaponized form of capitalist accumulation. Moreover, it requires the constant 'becoming-minor' of these subjects so as maintain this resistance. In other words, there is an emancipatory quality in acknowledging minor subjects' ability to radically nullify capitalist subjectivities. Deleuze and Guattari's notion of becoming-minor, as discussed above, necessitates being *of* the majority by destabilizing it, before fleeing to form new territories (or subjects) of minority; after all, 'you don't deviate from the majority unless there is a little detail that starts to swell and carries you off'.[12]

Exposing majoritarian thinking to those minor subjects with vastly different experiences of the world questions its perceived limits, forcing the destabilization of the built-up social conditions (and 'normalized' bodies) of male, straight, able-bodied whiteness. Moreover it exposes the injustices, discrimination, marginalization and everyday violence that normalization into a capitalist discourse can bring about. Therefore, the ethic of minoritarianism encourages embracing these alternative experiences as fully as possible and uses them to forge a common world that more people can shape.

It encourages an empathetic connection, one that triggers the mirror neurons to inculcate those same experiences in yourself (see Ethic 1). There is a myriad of unconscious biases that play out on a daily basis when people are confronted with disabled subjects. They judge the parents whose (autistic) child is having a tantrum in public; people stop trying to communicate with someone when they find out they're deaf; shoppers will deliberately use the counter not staffed by the person with cerebral palsy. There are many other ways in which majoritarian practices refuse the experiences of these marginalized people because of ritualized behaviours, social barriers, guilt, time pressures, fear of external judgement and so on.

Here, we can return to the pertinent work of Judith Butler. Critiquing Louis Althusser's notion of interpellation[13] and how it dominates contemporary theories of subject formation, she argues that becoming-subject requires a submission to the 'law', just as much as it does the law's power of subjection. In an Althusserian view of becoming, Butler argues that 'a subject is hailed, the subject turns around, and the subject then accepts the terms by which he or she is hailed'.[14] In other words, there is 'guilt' on the part of all of us that makes us conform to a particular subject that capitalism demands, be that major (a normalized body) or minor (a body that can be co-opted). We 'turn' (from constant becoming) when hailed by the state, capitalism or any other hegemony. Such a mindset, Butler argues, is predicated upon the view that it is better to 'be' minor and subjugated than not to 'be' at all. However, there are alterative possibilities.

> Such possibility would require a different kind of turn, one that, enabled by the law, turns away from the law, resisting its lure of identity, an agency that outruns and counters the conditions of its emergence. Such a turn demands a

Ethic 3: Minoritarianism

willingness *not* to be – a critical desubjectivation – in order to expose the law as less powerful than it seems.[15]

Hence, if we continue to reject 'being' in favour of 'desubjectivation', then the major/minor dualism breaks down, and capitalism loses its ability to articulate, and hence other or appropriate. Here, Butler is advocating a radical destabilization of identity so as to 'expose the law as less powerful than it seems'. So, in radically connecting with other minor identities, by refusing the hegemony of capitalist subjectivities, the major becomes porous and loses its perceived rigidity.

And this is how intersectionality aids in the ethical process of minoritarianism. When fugitive lines of marginality career into each other, the subsequent hybrid subjects that are created are new and alien to capitalist structures, and thus provide spaces for those marginal subjects to voice their concerns, create new protocols and, if lucky enough, celebrate their difference. Think of the large-scale protest marches and rallies of Black Lives Matter and those that are occurring in response to climate change (e.g. the Extinction Rebellion protests). These events create spaces where people can convene, debate and perform their hitherto oppressed subjects. The Black Trans Lives Matter march in June 2020 in the US would not have been possible without the emancipatory energies that were released by the broader Black Lives Matter movement. The Deaf community started Black Deaf Lives Matter in reaction to the lack of sign language on the marches and protest rallies. In response, more movements all over the world have been using Black Deaf interpreters. Minoritarianism breeds further minoritarianism; it is ethically infectious.

What is more, these intersectionalities have the powerful potential to radically alter those socially constructed biases that have built up over time. So for those who are not considered disabled by mainstream political, commercial and/or social discourses, the ethic of minoritarianism encourages majority subjects to seek out alternative subjectivities that are radically different, to open up new relations that push back against the 'naturalized' desire to maintain a subjective distance. What is more, it encourages a desubjectivation (to use Butler's language) of the self, a resistance to majoritarian and minoritarian identities. For those people already occupying the minor ground, rather than waiting for co-option, minoritarianism encourages 'resisting the lure of identity' and maintaining destabilization by infiltrating majoritarian subjectivities with alternative experiences.

Minoritarianism

Minoritarianist thinking is an ethical factor in bringing about a planetary commons *precisely* because it allows for the opening up of spaces in the margins for commoning to occur. By first realizing how the conflicting pathways of prejudice create different forms of subjugation, then engaging in intersectional thinking, and finally, and most crucially, empathetically engaging with those who have 'other' forms of subjugation than your own, minoritarianism encapsulates ethical practice that spreads the desire for equality to ever more groups of people (and their non-human environments).

Deleuze and Guattari's vision of becoming-minor is critical in 'breaking free' of majoritarian thinking, but, as this Ethic has shown, the specificities of which realms these breaks flee *into* are just as important. As the environmental scientist and social theorist Cindi Katz has put it:

Ethic 3: Minoritarianism

> Deleuze and Guattari's vision of the minor is promising because it creates new forms of subjectivity (both for 'majors' and for 'minors'); it recognizes (and depends on) the agency of 'others' in precipitating crises and thus social transformation; it offers a theory of transformation that works from within a relationship of oppression; and it offers flexible means for thinking about practice in new and revitalized ways.[16]

Therefore, via transcending the major/minor subjective dialectic, an ethical process of minoritarianism is being practised. There are, of course, practical and political endeavours that can provide the structural changes necessary to catalyse this ethical behaviour. There are movements within educational institutions both at school and higher education level to decolonize syllabi. This involves changing what is taught at school to reflect the structural racism that built the countries of the global north, putting more Black faces on lecture slides and Black authors on reading lists, and using museums and cultural institutions to inform people about the violence that acquired the collections. Many prestigious museums from around the world are attempting (too slowly in many people's opinions) to educate and inform visitors about their colonial legacies and illegal acquisition of their treasures. Initiatives such as the 'uncomfortable art tours' conducted in London of the V&A, British Museum and National Portrait Gallery are a pointer to these movements.

Also, reparations were thrust into mainstream debate (within US society at least) when the *Atlantic* writer Ta-Nehisi Coates penned his now famous 'the case for reparations' essay in 2014,[17] which has seen them debated in the US Congress. The call for global reparations from colonial powers to their former colonies is also growing louder, and with climate change disproportionately affecting global south countries, a system of resource reallocation from north to south will be vital. Both the decolonialization and reparations arguments contain a minoritarian ethic in so far as they acknowledge the repressive regimes that the majority has imposed upon the minorities, and as such will be a vital policy tool in enacting a planetary commons.

Ethic 4: Decodification

During the start of the coronavirus pandemic in the spring of 2020, there was much debate over numbers. There was a myriad of statistics used to measure the spread of infection, the death toll, the amount of testing being done and even the number of items of personal protective equipment (PPE) being given to front-line medical and healthcare staff. There was contestation from various governments around the world (notably the US) that questioned whether the official number in the cauldron of the pandemic, China, reflected reality. Indeed, in response to this, on 17 April 2020, the city of the first outbreak, Wuhan, admitted that it had vastly underestimated its death toll and it was in fact 50 per cent higher than it had previously said.[1] In the UK, the debate around the numbers seemed to reach absurd levels of introspection. There was criticism of the government that its death toll only ever included those who died in hospital. The government then started to include deaths in care homes, but those in the community were still left uncounted. Then there was the furore around the number of tests being conducted. On 1 April, Matt Hancock, the UK Secretary of State for Health and Social Care, promised that the government would be testing 100,000 people by the end of the month. As 30 April drew ever closer, the public debate about whether or not these numbers would be reached became more intense than any questioning of whether or not the test would help stem the spread of the virus. On the day of reckoning itself, the government triumphantly declared at the daily press briefing that the figure had been reached, although critics argued – with some credible evidence[2] – that the figure had been fiddled, with tests that had only been mailed out in the post included. Essentially, the UK government did all it could to be able to *claim* it had carried out 100,000 tests, without actually confirming whether it had done the tests or not. A few weeks later, it stopped publishing the figure altogether, seemingly abandoning any attempt to claim it had reached a target.

In setting a figure, and then manipulating the definitions and data to have seemingly reached it, the UK government was able to construct a narrative of success to deflect the growing criticism of its response to the coronavirus crisis. Essentially, the numbers were used as a public relations tool, something that aided the narrative that the government wanted to tell. The analytical capability to quantitatively measure testing as a means to prevent further infections was somehow lost to the figure's representative power as counter-narrative to the criticism that had been levelled at the government.

Conversely, when the UK became the country with one of the highest death counts in the world (only being 'beaten' by Trump's United States of America and Bolsanaro's Brazil), suddenly the government media line was not to be concerned with figures and statistics at this time. It seemed that the numbers were important to champion only when they portrayed the narrative that the government wanted. In essence, it was the public relations story that mattered,

Ethic 4: Decodification

not the numbers. There is evidence-based decision-making, but this was decision-based evidence-making.

And herein lies the problem of numbers, statistical representation and codification more broadly: it is too often used by capitalist hegemons as a representation of a constructed unreality that they impose upon society to tell a story that they want to tell. This is a fundamental part of neoliberal governmentality. Indeed, it has been argued that 'neoliberalism was born out of projects of world observation, global statistics gathering, and international investigations of the business cycle'.[3] The common language of neoliberalism is statistics. So by using numbers to codify more and more of the *un*codified world, neoliberalism has rendered reality more amenable to computation, and therefore applied an exchange value, and therefore made it ripe for capitalist accumulation. The ethic of *decodification* therefore attempts to resist this process.

To explain decodification it is first important to outline exactly what codification is. Second, knowing how the world is shunted and nudged into a codified landscape is critical because capitalist society moulds it into a flattened plane of comparability so everything can be ranked, ordered, and assigned a value that can be exchanged with everything else. Third – and this is where the *de*codification happens – there are groups of people, communities, inventions and ideologies resisting this process. It is to each of these three themes that I will now turn.

Codification: a philosophy of numbers

What do I mean when I say 'codification'? Put simply, it is the 'coding' of anything and everything in the world so as to assign it a specific *value*, namely an exchange value in the strict Marxist tradition. In other words, codification is a deliberate process or manoeuvre by those invested in capitalism's growth to force an exchangeable 'value' on something that does not have one yet. Codification is therefore more akin to a process of 'primitive accumulation', which involves a piece of common land (for example) being commandeered by a private force, the residents ejected, and the land carved up for privatization and profit.[4] However instead of abrupt enclosure of a once common resource, codification as a process is more about 'prepping' tacit phenomena (which do not have to be necessarily common) for privatized exchange.

The term is perhaps most readily used in legal language; to codify is to effectively write down a new law that will form part of a broader 'code' of legislation for a government, a company, a school, whatever the institution demands. Also, a 'code' is often a language that few can understand. A secret code or lines of computer code, for example, can only be read by those with the access or knowledge to do so. A code in these readings, then, acts as a go-to guide of what is permissible or not; a code of conduct, religious code or even a code of *ethics*. However, as was outlined in the introduction, this articulation of ethics is the predetermined kind that we want to move away from. They are stringent frameworks of behaviour that are often immutable and formulated within the confines of official or unofficial institutions (such a code could be attributed to a national government and an illicit group of dark net hackers in equal measure). Hence codification not only preps for capitalist exchange, but in doing so, it 'flattens' qualitative difference.

For example, university league tables have codified the once public and common resource of education and the pursuit of knowledge. They have created an arena in which

Ethic 4: Decodification

all universities can be ranked and judged on particular variables that force variegated processes that have very different patterns of behaviour into single, blunt and clunky metrics (e.g. graduate earnings, student satisfaction scores, research excellence and so on – each of them having very different kinds of qualities, *raisons d'être* and nuances across different universities, yet all measured in the same way). These league tables, while producing a hierarchy of institutions, have actually 'flattened' the world of university education and reduced its visibility in this world to a set number of variables that can be attributed across every university around the globe. This has created a 'marketplace' of higher education, where potential students, or consumers, can now have a supposedly transparent and informed knowledge about every degree in every institution. It is the classic case of marketization, and it has been going on in universities for decades. We see similar processes applied to cities, in which 'quality of life' variables such as pollution, access to green space, educational attainment and income are indexed and ranked, and individual cities are cumulatively scored. Again, the essence of urbanity or what it means to be a 'good' or 'bad' city – fundamental questions that have beset social science scholars for centuries – is seemingly solved with the quick infusion of some statistics and quantification about the 'quality' of city life.[5]

These seemingly neat and uncomplicated hierarchies of what are very qualitatively different entities (in this case universities and cities, but there are many others) are a symptom of *codification*: it erases these qualitative differences in favour of a forced compartmentalization using blunt instruments of statistical equivalence and metrics. The advocates of such measures will point to the neutralizing force of quantification; statistics offer transparency and a way to equalize different phenomena so to make them analysable, programmable and therefore malleable. In other words, quantitative analysis provides a way to easily assess a socio-cultural realm and offer ways of bettering it via interventions.

But are statistics actually that reliable? Are the numbers as exact as they seem? Take the expression of one third as a percentage. In school, many of us were amazed that when given the sum of dividing 100 by 3, we were left with a recurring remainder, a '3' that went on and on and on and on. The realization of the 'remainder' is etched on my memory; I could tell you exactly where I was, what I was wearing and the friends I was with. How could the neatness of numbers that I had hitherto been accustomed to now fail so spectacularly? Perhaps it was the start of my distrust of the quantitative as a means of accurate worldly analysis. Because the expression of a third as a percentage never ends; no matter how many 3s I put into it, there will always be a remainder; so does 33.333333 . . . % equal a third?

Mathematical philosophy has argued this point – whether 0.999999 . . . = 1 or not – for centuries. The endless streams of equations, the proof (or disproof) by geometrics, by pre-calculus or by cultural metaphor I will spare you. Ultimately, as the mathematician G. H. Hardy argued, it comes down to whether you choose to define 0.999999 . . . as equal to 1 in order to solve another problem (without creating another one elsewhere), or to keep going to the next 9, and then the next and the next, to see how far you can go.[6] In other words:

> We're untroubled by the fact that the English language sometimes uses two different strings of letters (i.e. two words) to refer synonymously to the same thing in the world. In the same way, it's not so bad that two different strings of digits can refer to the same number. Is 0.999...

Ethic 4: Decodification

equal to 1? It is, but only because we've *collectively decided* that 1 is the right thing for that repeated decimal to mean.[7]

It is the 'collectively decided' bit that exposes the constructed nature of this mathematical reality. Over the centuries, mathematicians (and those in related fields) have debated, discussed, deliberated and disputed over whether 0.999999... = 1. But in the interests of moving the discipline forward and being able to functionally use 1 as analogous to 0.999999... for other more pressing and practical mathematical, physical and engineering problems, there was a collective consensus, forged over time, through peer review and debated discussion within and around the discipline, to do away with the errant infinite remainder.

But despite this consensus, from a purely statistical point of view, ambiguity remains; it has just been overlooked qualitatively. The fact remains that 0.99999... doesn't = 1, and hence numbers resolutely fail to account for reality; it is only through a collective qualitative reality that 0.99999... = 1. Hence numbers are fallible. They are illusionary. They are mischievous. The geographer Marcus Doel goes even further to express that he detests every one. 'Not one in particular: just one in general.'[8] The continual count of one plus one plus one, he argues, is a violent act on the world. It 'liquidates heterogeneity' and reduces 'ambivalence ... to equivalence'.[9] In the way rankings distil universities and cities down to their measurable metrics, numbers are the ultimate mechanism of distillation. The One, as Doel remarks, is the universal arbiter to which everything can be reduced. So while Freddie Mercury in 1986 sang of there being 'one flesh, one bone, one true religion, one voice, one hope, one real decision', they may all be virtuous and different, but they are still all only *one*. They are still all the same.

Numbers, as Doel notes, are violently arbitrary in that they eradicate difference. The practice of statistical analysis, therefore, can be thought of as a violent discipline if conducted in hegemonic ways. The anthropologist Sally Engle Merry has argued that statistics which are not grounded and contextualized in local knowledges and practices risk producing information (as opposed to understanding) that is partial, misleading and, at its most extreme, dangerous. She used examples from the global research into human trafficking and domestic abuse. In doing so, she found that indices and statistics that are produced from initial studies that are more grounded, qualitative and from those who are the subject of the research give more reliable and meaningful results. For example, in analysing the treatment of abused women in the US courts, she found that a qualitative, interview- and focus-group-based study conducted by the suffering women themselves, as opposed to a global UN study that used secondary information from legal documents, produced results that were far more reflective of the situation. This study was able to produce statistical *knowledge* that was based on social and local geographical contexts, rather than *information* based on a generalized search for already-available statistics.[10] And it is this distinction between information and knowledge which is the violence of numbers. Statistical analysis can either 'flatten' the world into informational data that can be harvested, or produce knowledge that illuminates the differences. Merry argued that 'those who create indicators aspire to measure the world but, in practice, create the world they are measuring'.[11] Therefore it is worth repeating that there is evidence-based decision-making, but also decision-based evidence-making. For Merry, there is nothing inherently wrong about statistical analyses, as long as they bring with them the qualitative worlds on which they are built.

Ethic 4: Decodification

Merry is therefore making a broader distinction between qualitative and quantitative analysis, and the powers that they wield in the public realm. She goes onto argue that '[The] social aspect of indicators is typically ignored in the face of trust in numbers, cultural assumptions about the objectivity of numbers and the value of technical rationality.'[12] This point is an important one for the ethic of decodification, because it is in that 'social aspect' – the qualitative realities that are lost in the codification of the world – where the commons is to be found and nurtured. To find it, though, is not easy, not least because our world has been saturated with codifying technologies. We live in a world in which the codification – and therefore *de*socializing – of life happens via the ever-deepening embeddedness of technology in our daily lives. To decodify society towards a planetary commons, then, it is useful to know how to first decodify ourselves.

The quantified self

Codification in the contemporary digitized world can be very personal and even psychological. This is a symptom of the broader 'quantified self', a cultural 'movement' first given a name in 2007 by Gary Wolf and Kevin Kelly, a couple of technology journalists. Yoked very much to the march of personal technologies, the quantified self is the management of the body, in increasingly fine detail, via statistical operationalization.[13] The advent of wearable technology has seen the datafication of many once unrecorded human behavioural and psychological processes. A humble smart watch can now measure the effectiveness of our heart rate, sleep and even the depth of our breathing. It has also influenced governmental policy; for example, the UK NHS encourages people to eat five portions of fruit and vegetables a day, as well as do 10,000 steps (provided you have a digital pedometer, of course).

Much of this is important within healthcare practice and policy; we are often hooked up to heart monitors or pulse oximeters while in hospital in order to diagnose or cure an illness. And government-led pubic health programmes are nothing new. But now, because these technologies are small and cheap enough to fit inside a smart device, they have become widespread, and gone beyond purely diagnostic functions to form part of the biopolitics of neoliberal life. They are used performatively to curate our daily routine. Often sold on the premise of 'improving well-being' (itself a distinctly neoliberal vernacular designed to catalyse the ideology of the enterprising self,[14] discussed further in the next Ethic of slowness), these technologies imprint quantification on more aspects of our life. But beyond simply quantifying our daily lives (via how many steps we do, how regular our heart rate is or how long our REM sleep cycle is), these technologies are *codifying* our bodies in the same way that universities or cities are codified so as to allow marketization. The quantified self 'movement' is flattening out the array of human practices and emotions into a level and, crucially, quantified playing field. In the process, we are being conditioned, by these technologies and the broader biopolitical framework of governmental control, to be more amenable to the accumulative and privatizing affects of capitalism.

This process was articulated by Deleuze in 1990, when he wrote a short essay, 'Postscript on the societies of control'. In signalling a move away from a Foucauldian understanding of societies of discipline, in which power was exercised by disciplinary institutions such as the prison, the hospital, the school and the factory, Deleuze was arguing that power

Ethic 4: Decodification

in contemporary society operates as a form of control. Now, Deleuze argued in 1990, power gives us freedom to do what we want to do; it's just that when we do it, we are offering up more ways of being controlled. He used the metaphor of a highway where we are free to drive around for as long as we like, but we have to keep going in one direction. It is easy to see how it relates to today's society when we take the humble smartphone. Almost everyone will have one, possibly more. Every time we use it we give up some kind of data. Location, food preference, social media likes, even our eye position on the screen: the smartphone channels our desire so those who want that data to effect more means of control (namely corporations and governments) can continue to obtain it. The streams of 1s and 0s that emanate from our phones, credit cards, travel cards, smart watches and even our smart condoms all go towards creating more forms of corporate, governmental and capitalistic control.

In Foucault's reading of the disciplinary society, the ability to administer power rested on two societal denominations: the signature and the number. The signature designated an individual within society, and the number administrated that person within a broader mass. This dual differentiation operated in tandem to keep the masses obedient and 'in check'. However, in signalling a move away from these Foucauldian disciplinary societies, Deleuze stated that 'what is important is no longer either a signature or a number, but a *code*: the code is a password'.[15] What Deleuze is arguing here is that the coding of society is not dualistic any more, but operates a version of control by softer forms of hegemonic power. This *code* therefore is a means by which we are free to act as we please, but in so doing, we are creating more data about ourselves that can be watched, bought, sold and traded.

Ever-increasing parts of our lives are being captured by this code and used to either monitor us or advertise to us. Deleuze goes on to argue that within societies of control, we are no longer individuals, but *dividuals*; through our engagement with the coding of the society of control, we have been divided up into constituent parts – our work, our leisure, our education, our legal standing: they are all parts of different systems of control that often overlap, but continue to monitor and affect us wherever we happen to be. As such, the exercise of power is no longer confined spatially to a factory; we take our work home with us. The exercise of power is no longer confined spatially to a school or university; we can learn on the go. The exercise of power is no longer confined spatially to a prison; we can wear tracking devices and be seen on CCTV. The exercise of power is no longer confined spatially to a hospital; we can see our GP on our phones and track our heartbeat with our watch. And all of these forms of power (and many more) are under the watchful eye of capitalist ideologues, finding new ways to monetize these forms of power.

And perhaps now we are even beyond Deleuze's vivid yet prescient imagination. Along with the digitization of life and our bodies, there is a chemical, physiological and biomolecular control method, due to the preponderance of what the philosopher Paul B. Preciado calls *pharmacopornographic* management of the body. *Pharmaco-* because of the mass consumption of antidepressants, anxiolytics and other forms of chemical interventions; and *-pornographic* because of 'the incitement of consumption and the constant production of a regulated and quantifiable pleasure'.[16] In the wake of Covid-19, Preciado argues that the governmental response itself has been pharmacopornographic: we have been reduced to mask-wearing (so faceless), digitally interfacing (via contactless payments), traceable (via

Ethic 4: Decodification

the various track-and-trace apps) and virtual (via endless online meetings) beings. We have become neoliberal actors par excellence. Preciado goes on to argue that we 'are not physical agents but rather tele-producers; [we] are *codes*, pixels, bank accounts, doors without names, addresses to which Amazon can send its orders'.[17] The control that capitalist hegemony has over the human biome has multiple pathways that ensnare us in a Gordian knot of biopolitical measures.

Via these societies of control, be they digital and/or pharmacopornographic, we have been *coded* into a capitalistic society. Hence why, in order to resist capitalistic control and create a different form of society that is held in common, there needs to be a *de*codification. In turning away from this process of coding, and by creating systems that allow people to plug *in* to benefit, but to then plug *out* so as not to hand over control, we are being ethical towards the realization of a planetary commons. This is because plugging out curtails the appropriative mechanisms of capitalism that shut down the commoning process.

Luckily, there are plenty of counter-movements to the 'quantified self' that highlight how there are indeed ways to plug out of a system of codified technological determinism that enables capitalism to infiltrate our psychology, emotions and physiology. There are ways that technology works to allow more control over our humanity, rather than surrendering more of it to the accumulative discourses of capitalism; there are techniques of bodily decodification.

Bodily decodification

The compass was not originally invented for navigation. In the second century BCE, the Chinese (during the Han

Dynasty) used pieces of lodestone, a naturally magnetic rock, suspended in the air to find out where best to place houses, plough fields and also search for other precious metals. It was a form of geomancy that we now know as *feng shui*.[18] It wasn't for another millennium that lodestone was used to create the compass as a navigational tool. It is theorized that through the use of lodestone to situate their living and working arrangements, the ancient Chinese were able to attune their own situational senses to 'know' where magnetic north was. In effect, they had an extra directional sense to complement the five traditional human senses (a sense that migratory birds utilize on an annual basis with immense precision). Over the course of time, that sense has been lost. Technology has replaced our navigational sense, and with the embedding of mapping technology and geostationary satellites that track our movements, our directional sensing has been completely outsourced to machines.

However, some people are attempting to wrest this skill back into human hands, or at least, into the human body. Researchers in Germany created a 'sense belt' that is worn around the waist, and vibrates whenever magnetic north is faced. The researchers performed a longitudinal study in which participants wore this belt continuously for seven weeks. Almost all participants, after taking the belt off, were able to 'remember' (via muscle memory) where magnetic north was. The results of the experiment suggest that 'new sensorimotor contingencies can be acquired using sensory augmentation'.[19] In other words, using this piece of technology but, crucially, *taking it off afterwards*, the human body can be augmented with a so-called 'supersense'. This has proven extremely useful for people with blindness, as their ability to navigate has been greatly enhanced, given that previously they would have had to

Ethic 4: Decodification

use spoken directions or hand-held devices. Now, they can use the new-found directional 'sense' to aid their mobility in the modern world.

There are many, far more extreme examples of people augmenting their bodies to achieve extra-sensory perception, but what is important to note here is that so-called 'hyperhumanism'[20] is the act of using technology to enhance the already-existing, perhaps latent but no less present, capabilities of the human body. It is making technology 'work' for us and not the other way round. And in so doing, it is limiting the codification of human lives into the digital streams of social media and smartphone addiction. More than that, it is an active decodification. This is a step beyond a simple 'digital detox', a technique where we are encouraged (often by self-help gurus looking for a consultancy fee) to go for a period of time without any smart devices in our lives; no access to social media, streaming services or computer games. For sure, there is some merit in a digital detox, but the evidence that a period of abstinence from smart devices increases well-being is shaky to say the least.[21] Decodification is a step beyond this. It starts with the individual being able to extricate their lives from codifying technology, but in order for the ethical commitment of a planetary commons to be realized, it must go beyond the individual and engender a *collective* action of decodification. After all, a planetary commons is predicated upon an interaction of the community (at whatever scale that is defined, be that a family, neighbourhood, region or planet) with its common resources; but that the community is first decodified, or at least striving towards decodification, is crucial.

Seven Ethics Against Capitalism

Urban decodification

So how can this be done? How can we scale up decodification from the self to the planetary? And what about the various scales – the community, subculture, neighbourhood, city, nation – in between? It is worth returning to Deleuze's seminal essay, 'Postscript on the societies of control', in which he recounts the words of his friend Félix Guattari;

> Imagine, if you will, a city where one would be able to leave one's apartment, one's street, one's neighborhood, thanks to one's (dividual) electronic card that raises a given barrier; but the card could just as easily be rejected on a given day or between certain hours; what counts is not the barrier but the computer that tracks each person's position – licit or illicit – and effects a universal modulation.[22]

Now compare that with a quote from Thad Sheely, who was a mid-level manager within the new Hudson Yards 'smart city' project in New York City from 2014, about how the new city will use residents' data: 'Basically, we'll be the funnel and collect the data so they can put [the information] though their spin cycle in the cloud, and then provide an interface for us to be able to access the information.'[23] Notwithstanding the rather nugatory phraseology of 'spin cycle in the cloud', there is a haunting resonance between the two quotes that evidences the actuality of the 'societies of control' in the modern-day smart city. And so, to aid in a decodification ethic on a planetary scale, it is worth looking at how decodification can happen – and is happening – on the city and neighbourhood scale.

Hudson Yards has been one of New York's most ambitious building programmes, not only because of

Ethic 4: Decodification

the engineering practices that it had to employ to build skyscrapers above a working train yard, but also in the neoliberal practices of contemporary smart city design it employs. Urban scholar Shannon Mattern has described it as 'a clean, efficient urban machine; a carefully curated cultural experience ... a harmonious community that behaves in accordance with the rules; a city that plays by the numbers'.[24] It is New York's newest neighbourhood, but one that has been completely constructed under the rubric of 'datafication', where the input, output and processing of data has been first and foremost in the city's design. Residents are invited (or perhaps forced) to use biometric data when entering their building or depositing their trash down the chute. Surveillance and data-gathering technologies permeate the architecture. The site is also host to publicly accessible spaces such as an upmarket shopping mall and, famously, the 'Vessel' – a giant piece of public art by Thomas Heatherwick. Consisting of interlocking staircases and escalators, it reaches 150 feet into the air and cost upwards of a reported $200m. One of the main controversies around the site (other than the obvious questions around why this shiny, monochromatic monument to the vacuity of contemporary commissioned urban 'art' was allowed to exist in the first place) was buried deep in the terms and conditions of entering it (which is free, but ticketed). Anyone taking a photo of the 'Vessel' was automatically letting the owners use that photo for 'any purpose whatsoever in any and all media (in either case, now known or developed later)'. After a public outcry, this condition was removed, but the fact that it even existed in the first place speaks to how prized data is by private stakeholders in the city.

Across the city from Hudson Yards, in Long Island City, Queens, Amazon was busy planning for 'HQ2', a site that would house the company's 'second' headquarters. The site was chosen after a lengthy and very public national competition in which many places were criticized for offering generous tax breaks to Amazon to locate in their city. Eventually, in 2017, Long Island City was chosen as the winning site. However, the locals were distinctly unimpressed and began to campaign against the decision, claiming that Amazon was an anti-union corporation and the gentrifying forces of such a project would be devastating to the area. In February 2019, only three months after the project was announced, Amazon pulled out.

A similar process happened in Berlin, when in 2018 the 'Fuck off Google' campaign was successful in stopping the internet search giant from opening a 'smart campus' in the Kreuzberg area of the city. And in May 2020, Sidewalk Labs – the smart city building partner of Google, which had successfully implemented Hudson Yards – cancelled their supposedly technologically groundbreaking Toronto project, claiming the coronavirus pandemic made it economically unfeasible (although here too, they had faced years of protests).[25]

All these (admittedly global north) examples could be explained away as the 'normal' forms of the anti-gentrification campaigns that have been witnessed across the global cities of the world. But the fact that it is Silicon Valley corporations that are targeted rather than 'traditional' real estate companies speaks to the embeddedness of technological datafication in the fabric of contemporary urban development. The large-scale protests against these new smart city programmes were predicated upon what the protesters saw as gentrifying effects that come in the guise of higher rents, displacement of the urban poor (who are often working-class and of ethnic minorities) and homogenization of the urban landscape. But within

Ethic 4: Decodification

the protests was also a call to recognize the ways in which citizens are being forced to give up more and more personal and domestic data in the service of the large private technology firms.

The residents, workers and even visitors to these smart cities are engaging in 'smart citizenship',[26] in which giving up data on yourself and the environment around you becomes part of 'good citizenry'; any attempt to resist or conceal yourself suddenly becomes cause for suspicion. This clearly has issues around the erosion of civil liberties and the privatization of urban space,[27] but also, importantly to the arguments of this ethic, it smothers the city – and our citizenship of it – in a layer of data that first has to be produced, then mined. The smart city, then, is more than the use of technology to make urban infrastructure more efficient and increase privatization; it is the codification of urban citizenship into another form of control. In the same way that the quantified self fractures the nature of individuality into constituent realms of codified life, the smart city movement reduces what it means to be a city into different data streams that can be codified and, importantly, capitalized upon.

Moreover, the very act of citizenship itself – the democratic engagement with the city, its governing institutions and fellow citizens – is being codified via the ubiquity of technological interactions with the urban realm. How we get around the city, order food, shop and find love are computed through what geographer James Ash has described as the 'interface envelope'; the enfolding of the immediate and localized time and space around us by ambient technologies to morph our perceptions and our bodily functions to maximize profitability.[28] In 2016, the filmmaker Keiichi Matsuda produced *Hyper-Reality*, a futuristic representation of just such an interface envelope.[29] It depicts a somewhat dystopian world where our field of vision is layered with multiple augmented realities that cover the city with media to swipe, pinch and zoom. The short film features a woman, clearly bored with the shallowness of the digitally immersive world she occupies. She commutes, shops and vies for more 'loyalty points' from the paragons of commercial dominance that control and mediate the interaction with the physical city around her. In one telling scene in which she resets her augmented vision, the physical reality of the supermarket she is in is suddenly exposed: the monochrome aisles, screaming baby and sea of bland QR codes. When I show this film to my students, they are amazed at the fantasy city it depicts. But I then go on to talk about China's social credit system, which works along eerily similar lines, and suddenly they become far more nervous about its near-reality. The codified smart city, therefore, is a neoliberal utopia where self-interest is rife, citizens become consumers and the landscape is devoid of any collective democratic commonality.

Planetary decodification

But it does not have to be this way. There are many urban neighbourhoods and communities that are using the immersion in technology and the 'big data' it produces to forge a commoning of the city rather than an individualization of it. They are using the emancipatory potential of socialized technology to identify and suggest changes in the urban landscape. For example, in the Netherlands, Verbeterdebuurt is a public technology whereby citizens can suggest ideas on how to improve their city, start a signatory list and contact the local government direct to obtain funds.[30] In the US, AirCasting is an open-source

Ethic 4: Decodification

environmental data-mapping platform where citizens can measure and upload real-time pollution levels, thereby providing data to use for community-led cleaner-air lobbying campaigns. FixMyStreet is a UK-based online portal where citizens can upload information about their location that the local council needs to act upon. During the coronavirus pandemic, mutual aid networks were mapped using open-source software; during the Black Lives Matter protests of 2020, apps were used to alert protestors to police movements, and open-source education software was created to school people on the racist symbolism of statues. There is a whole range of community- or activist-led data platforms that have been used to facilitate a more democratic, activist or even anti-capitalist movement. They are all examples that use a community-driven platform that is designed to open up the commoning and socialized potential of the technology rather than to *dividualize* us (to continue with the Deleuzian language) into codified realms of capitalistic computation.

And these 'digital commons' platforms (which are often funded from public money, or charity- or crowd-funded) do this by shifting the power dynamic away from corporations and draconian government departments looking to codify the different parts of our lives, towards increasing the agency of the community (be that a neighbourhood, a city or an entire global population). Yes, they involve the continuing use of technology, but they are changing the fundamental use of that technology by facilitating a democratic engagement with the city and its governance. They bring the qualitative and community agency into the quantified encounter. Moreover, they have the potential to blur the lines between the city and who controls it. Currently, too much of the 'big data' that the city generates is automatically owned by private corporations (as in the Hudson Yards social media controversy) looking to instantly profile us for surveillance or commercialization purposes; it is, as I have argued throughout this Ethic, a codification process.

Hence, to decodify this data would be to wrest it away from these corporations, allow it to be generated by and for public use. Furthermore, the digitized nature of this commoning can boost its planetary potential. In an interview during the pandemic, the French philosopher Bruno Latour suggested that the virus showed us how to 'viralize' the world; individuals may feel small (something which the capitalist codification catalyses), but the virus showed us how ideas and empathy can spread around the world.[31] Hence these digital tools have the potential to radically connect all life around the planet in a commoning praxis if their productive rationale is shifted from corporate growth to democratic engagement. With these ideas in mind, then, what would an urbanized public transportation system look like that was able to use the platform and the reach of Uber? How much more would the vulnerable (during a pandemic or otherwise) be served by a commonly owned delivery platform the size of Amazon? These are not utopian questions, because these things are achievable if only a different, less capitalist and more common, approach is taken. The 'digital commons' is an already-existing part our world, and in a post-Covid society in which online communication, digital tracking of disease carriers and worries over data privacy are only going to increase, having this data in the hands of publicly owned institutions will be an important step in dismantling big tech's stranglehold on our codified lives.[32]

There is even a model right under our noses: Wikipedia (and indeed other commons-based online software such as Mozilla Firefox and Gnutella). The online encyclo-

Ethic 4: Decodification

paedia that we all take for granted is one such (largely) democratically managed, open-source, crowd-funded, live common resource. The data generated by the various editors and contributors from all over the planet is mediated and checked by everyone else, and the democratic implementation means it is one of the world's most trusted information portals. It is not completely horizontalized; it does contain elements of hierarchy in terms of the various levels of editors and its most reliable users (not to mention the influence of the founders Larry Sanger and Jimmy Wales). However, despite this, on a day-to-day basis, it is a digital resource managed by the planetary community that feeds off it. It is a community and a common resource in constant communication with each other.[33]

The community-based apps, encyclopaedias and citizen engagements with smart infrastructure around the city *humanize* the digital commons, give it a democratic purpose and make it less amenable to any profiteering exercise. They *de*codify. How much bigger could they get with the right political will to empower them? I will leave the last word to Paul B. Preciado, who has unequivocally argued for using the lockdown of the Covid-19 pandemic to perform a radical decodification and uncoupling from the society of control. He writes: 'Let us turn off our cell phones, let us disconnect from the internet. Let us stage a big blackout against the satellites observing us, and let us consider the coming revolution together.'[34]

Ethic 5: Slowness

There is an inherent paradox within capitalism: it acts slowly by speeding up. In cranking up the speed at which consumers buy, discard and rebuy goods and services, its longer-term violent effects of accumulation by dispossession are achieved. The long progress of capitalism has been predicated upon a technologically deterministic process that has seen horse-drawn canal boats, the steam engine, refrigeration, just-in-time production, plastic packaging, the internet, robot factory workers, the smartphone, artificial intelligence and shopping algorithms increase the pace at which natural resources can be extracted from the commons and transformed into goods we can consume and throw away. The slow grind of transforming us into consumers of the ever-increasingly commodified material world all around us requires the speeding up of the consumption process. To enact its slow violence,[1] capitalism requires us to think quickly.

In their magisterially rich theoretical work *The New Spirit of Capitalism*, the sociologists Luc Boltanski and Eve Chiapello perform a vivid and complete dissection of the intricacies of modern capitalism. Taking their lead from some of the in-depth theoretical theses of the past, includ-

Ethic 5: Slowness

ing Marx's *Capital*, of course Weber's *The Protestant Ethic and the Spirit of Capitalism*, as well as Deleuze and Guattari's *Capitalism and Schizophrenia*, Boltanski and Chiapello show capitalism's growth is ultimately predicated upon our *desire*. This word brings with it complicated philosophies too rich to decipher here; suffice to say that our *desire* is what produces the worlds in which we live. Deleuze and Guattari call it *desire-production*, such is the power of innate human desire to create change in the being-in-a-world. It comes in the form of collective desire as well as individual; it creates connections, publics, families, commonalities and the commons. This desire can be tangible or completely tacit, but it is a presence that, when experienced inside or outside of the commodity sphere, comes in the form of enjoyment. Beyond merely enjoying a meal or the smell of a new car, desire can be enjoyment that is the collective sociality we have as humans. We can feel it (albeit perhaps in glimpses) when part of a mutual aid project, when engaging in spiritual worship, when dancing in a crowd or even at a football match when our team scores. It is a *force* that pushes at the unstable boundaries of our current understanding of being, and creates new territories and worlds that are, as yet, untouched by capital.

Yet, over the centuries, capitalism has perfected the mechanisms of controlling and channelling that desire to feed its own growth and stop us from creating these new worlds in the first place. A key part of that is the individualism that was explained in Ethic 1. But in addition Boltanski and Chiapello explain how, post-1968 and the cultural revolutions that came with it, capitalism reacted and now harnesses the revolutionary celebration of cultural differences and diversity to weaponize our own desire against us; it offers us speedy excitement and near-instantaneous pleasure via the commodity, on top of, or instead of, the stability and security via waged labour that it offered before.

In detailing how the 'spirit' of capitalism, then, maintains its existence post-1968, Boltanski and Chiapello explicate how excitement is used to channel fundamental human beings' desire (which these writers articulate as insatiability) not for the common good, but for sustaining capitalism over the long term. 'The spirit of capitalism ... activates insatiability in the form of excitement and liberation, while tethering it to moral exigencies that are going to restrict it by bringing the constraints of the common good to bear on it.'[2] Our collective desire to produce worlds beyond capitalism (knowingly or not, willingly or not) is therefore redirected by the spirit of existing capitalism via excitement and liberation, and through the rapid consumption of goods and participating in the rapidity of the marketplace. Excitement within the commodity sphere is instantaneous, and replaces a more ethereal *longue durée* form of enjoyment beyond it. Via the society of the spectacle[3] that capitalism has created, we are blind to any other form of enjoyment beyond consuming it instantly. We are not given a chance to think about desire's broader human and material connections.

Critical to the foundations of this ethic of slowness, then, is that capitalism's offer of excitement is invariably yoked to the temporal. Excitement via consumption is short-term; the enjoyment of desire-production is long-term. Boltanski and Chiapello's work posits capitalism as therefore manipulating the time frames of how desire operates in order to limit its transformative potential (beyond any transformation that would see capitalism diminish). Desire (or insatiability) is critical for capitalism to grow, hence the need for a 'spirit' to guide our desires; and one

Ethic 5: Slowness

sure-fire mechanism to do this is to limit and curtail excitement to short-term fixes.

But the desire-production that capitalism channels is critical to sustaining a planetary commons. Ethic 2 detailed how a transmateriality is vital in reconstituting the balance between the human and the nonhuman world, a major part of which is the realization of the spatial, social and temporal nuances of the commodity's production process. Throwaway culture is a brake on transmateriality because it relegates the thing-power of objects. Similarly with this ethic of slowness; throwaway culture is part of the process by which the spirit of capitalism offers excitement via the commodity.

The ethic of slowness, then, is to counter capitalism's version of excitement. It is to reject the rapidity of the commodity sphere in favour of thinking slow. Capitalism denies us the opportunity to dwell, to contemplate, to simply be; or at least any attempt at this is now hyper-commodified.[4]

But how do we slow down capitalism that, like a juggernaut going downhill with no brakes, seemingly cannot be stopped and is heading for disaster? First this ethic will detail how capitalism has perfected the science of pleasure as something that can be codified and therefore profited from. But as much as capitalism encourages us to speed up, there are also other ways within it that can aid us in slowing down. Movements such as slow media and slow fashion are important parts of resisting capitalism's constant need for quick returns, and while they can be part of capitalism via just another form of consumption, they offer kernels of commoning which can be expanded upon. Within the cracks of capitalism, then, there are moments of slowness that, if exposed and expanded, can begin to shake the foundations of capitalism built on the speed of financialization.

Happiness

In 2016, the federal government of the United Arab Emirates (UAE) launched the National Programme for Happiness and Wellbeing.[5] Within this ambitious plan was the creation of a 'National Charter for Happiness' which was designed to 'promote virtues of positive lifestyle in the community and a plan for the development of a happiness index to measure people's satisfaction'.[6] The country created multiple CEOs (chief executive officers) for happiness and well-being at various government levels, created a 'customer happiness formula', and initiated a happiness research institute, a well-being academy and a whole host of other initiatives aimed to promote the nebulous emotional state of happiness among its citizens. The result? The UAE was ranked the nineteenth-happiest nation in the world in 2020, measured by variables such as the weather, green spaces and access to water.

Notwithstanding the reduction of happiness to a simple case of having access to natural amenities, the fact that a globalized industry has been built up around the emotion of happiness is a critical feature of neoliberal capitalism. The sociologist William Davies, who wrote *The Happiness Industry* in 2015, argued that in line with the progression of neoliberalism, happiness has become the latest victim of capitalism's accumulative drive. Advances in personal technological capabilities that have codified and quantified our psychological capacities (as detailed in the previous Ethic) have allowed our emotions to be subject to the whims of capitalistic accumulation. Davies argues that despite all of us experiencing it otherwise, happiness is now a measurable phenomenon. Chemically via dopamine or serotonin, psychologically via flow or passive activity,

Ethic 5: Slowness

financially via how much money we have, geographically via how close we live to a green space: how happy we are as individuals (and therefore aggregately as a collective) can be quantified, measured, codified, ranked, packaged and sold.

Enjoyment itself is therefore part of neoliberal machinery; if we feel unhappy, we look to the commodity and/or pharmaceutical market to rectify the situation. However, linking the visceral intangibility of 'enjoyment' to the realm of capitalist consumption makes it subject to all the other unethical characteristics that come with that. It individualizes happiness, thereby neglecting socialized and collective enjoyment; it encourages us to consume the nonhuman material world to increase happiness, thereby negating its vitality; it demands we quantify it and, critically for this ethic, demand it *now*.

A truism that surrounds this debate about the codification of happiness is the difference between short-term pleasure and long-term enjoyment. The medical version will suggest that it is to do with chemicals in the brain. Dopamine, despite being the so-called 'pleasure' chemical, actually only drives motivation to do particular actions, be they pleasurable or not. It is a motivator, but it can motivate us to become addicted to harmful substances just as much as to 'pleasurable' ones. It is no secret any more that the social media giants of Silicon Valley spend millions upon millions of dollars in replicating the 'casino effect' on their products (so every time you scroll down to refresh you're essentially replicating the addict gambler at a slot machine[7]) in order to get the dopamine flowing in our brains.

Dopamine contrasts with serotonin, the chemical which supposedly regulates our longer-term mood. A synthetic version is often prescribed for those with depression, given that it has a proven ability to improve general well-being in the longer term. Psychologists will attribute the release of serotonin to 'flow activities', that is, those that require in-depth long-term haptic or bodily concentration such as exercise, writing, art and crafts, carpentry and the like.

The short-term/long-term duality, though, cannot be distilled to purely a chemical imbalance. If only it were that easy. It is more about desire in the Deleuzo-Guattarian sense and the imbrication of the individual in the social world around them. The short-termism that has come to quantify and commercialize our digitally mediated interactions with each other is built upon the broader structural pressures that maintain these instantaneous 'desire protocols'. In other words, capitalist short-termism cuts off our social connections (by forcing us to look to ourselves), makes us see material as just dead matter for us to consume, and codifies everything around us. Instantaneous gratification and the recalibration of pleasure are just some of the ways in which capitalism reprogrammes our desire away from creating alterative means of social interaction beyond marketized forms, and thereby maintains its own existence.

Despite the multifaceted and socialized way in which 'happiness' manifests itself, the physiological dichotomy between short-term hits of dopamine and long-term releases of serotonin has come to characterize the entire industry of happiness and how desire is controlled. The short-term 'dopaminic' hit can be manipulated so precisely by the media interfaces we are connected to (and the corporations that control them) that our very pleasure programming is being rewired around the instantaneous consumption that is now our constant demand of everything around us. The biopolitical control that capitalism has so effectively enacted upon us means that 'waiting'

Ethic 5: Slowness

or simply 'dwelling' becomes either a subversive act or a commodity in itself. The speed at which capitalism operates and the instantaneous gratification principle that it has implanted in us has created a near-perfect feedback loop that ensures its constant accumulation. By creating and then fulfilling short-term desire protocols all bundled up in one product, capitalism keeps us enveloped in digital interfaces, too impatient to look up at the decaying world around us. Patience really is a virtue that is increasingly becoming lost.

What this also means is that the longer-term enjoyment that can come from sustained engagement in creating a non-capitalist commons that is self-determining (such as an activist community, social movement, mutual aid project or self-contained eco-squat) is nullified. The codification and quantification of pleasure via the pharmacopornographic control of society require us to delve into the short-term hits of the market for happiness.

Resisting this, being *slow* ethically, is a process that deadens the accumulative actions of capitalism. Slowness in this sense defies all the violence of codification, and of commercialization more broadly, that comes with demanding rapidity in our desire. By simply slowing down as a collective (be that a community, a corporation, a government, a country or a global multitude), we can see a planetary commons come into view, because slowing down provides the space for us to see the benefit of all the other ethical commitments and can allow the increased ethical infectiousness of commoning. Being slow in making connections with others and the material world around us shows the use value (rather than the exchange value) it can engender.

But given the rapidity and hypermobility of capitalism in the twenty-first century, there are precious few examples of an ethical slowness, although they are increasing. There are a number of 'slow' movements and cultural genres that have a kernel of ethical truth, which can be applied to the creation of a planetary commons if taken seriously as a wider social characteristic. In the rest of this ethic, though, I would like to focus on two: slow media, and slow fashion.

Slow media

'The Ghan' is a passenger train service that runs north to south across the entirety of the Australian continent. From Adelaide in Southern Australia to Darwin in the Northern Territory, the track runs centrally via Alice Springs and takes 54 hours to travel the approximately 3,000 kilometres. Short for 'The Afghan Express', it is named after the Afghan camel trains that helped British explorers reach the interior of Australia. These days 'The Ghan' is a luxury train experience that takes travellers through some of the most desolate landscape on the planet. The miles of uninterrupted vistas and monotonous parallax views have become somewhat of a selling point; travel in luxury and marvel at the seemingly unlimited expanse that can bore and amaze simultaneously.

In 2018, an Australian television network broadcast a seventeen-hour[8] 'slow TV' programme that showcased the journey of 'The Ghan'. It used all the techniques that have long been the mainstay of the art house form of 'slow cinema'; long takes, very slow panning shots, fades that last up to a minute, no music so as to expose the raw soundscape, no dialogue, no real plot to speak of.[9] In many ways, slow cinema is anti-cinema. When writing on this most complex of art forms, Deleuze points to its

Ethic 5: Slowness

dislocating possibilities and that it can create 'any-instant-whatevers',[10] and Guy Debord says cinema is where 'détournement can attain its greatest effectiveness ... and its greatest beauty'.[11] Indeed, film theory scholars will talk of the oneiric qualities of film, in that cinema is dreamlike and transports us out of reality, to the other world that is disjointed, fleeting, and untethered to the steady process of 'real-world' time.

Slow cinema counters this rendition. It roots what is on the screen firmly in aesthetic reality with as little editing, cutting and quick camera movement as possible. More than that, it deliberately slows us down in two ways. It can either force us to lean into the screen analytically, scouring it for detail and marvelling at the marginalia (rather than focusing on the central motif of the shot), or it can be something that can simply wash over us, engendering almost a trance-like state. The seventeen-hour epic *The Ghan* has both of these in spades; it goes against all the entrenched tropes of cinema and jars the viewer into an unanalytical and trance-like stillness. One cannot help but be drawn into the landscape, the carriage interiors and the vanishing lines of the rail tracks. The train-journey-based slow TV genre emanated from Norway, which had a ten-hour show of a journey between Oslo and Bergen. Sitting in line with the broader Scandinavian cultures of *hygge* (roughly translating into cosiness), *lagom* (moderation) and *kalsarikänni* (which literally means 'pantsdrunk', to drink at home in your underwear, but more broadly signifying a form of carefree comfort), the slow TV genre has seeped into mainstream culture.

There is also a subculture of slow music. The American composer John Cage[12] wrote 'As Slow As Possible' for the organ or piano in 1985 but deliberately neglected to give the speed at which the piece should be played. After Cage's death, a group of philosophers, theologians and musicians met regularly in the 1990s to discuss just how slowly the song should be played. They came up with the answer that it should take 639 years. This is because one of the major innovations in musical technology, the Blockwerk organ, was completed in 1361, exactly 639 years prior to the turn of the millennium. A specialist organ was constructed in the very same place where the original was built back in 1361; a small rural convent in Halberstadt, Germany. The 'performance' was started on 5 September 2001 with seventeen months of silence. The first full chord change wasn't until 5 July 2008. Each movement lasts on average seventy-one years. That this piece of music is too long and too slow to ever be fully appreciated by any one individual, and indeed will need multiple generations to 'listen' to the whole piece, makes it a truly collective and fully socialized cultural artefact.

These slow cultural movements offer a different reality that has been usurped by the very *un*real hypermobility of modern-day capitalist life. *The Ghan*, 'As Slow As Possible' and other comparable slow pieces are of course couched in the capitalist production process, but within that, they carve out a space that is too often lost or deliberately curtailed by those power structures looking to keep us engaged in rapid consumption patterns. These slow forms offer a space (emotionally and temporally) for us to dwell, to contemplate, to meditate, to see the less immediate connections we have to the world, the people and things around us. This is a vital part of commoning precisely because this space is critical for mutualism, transmaterialism, minoritarianism and decodification to flourish. By learning from these pieces, by (in effect) practising how to be slow, commoning is something that can spread into other parts of our lives, our communities and society.

Ethic 5: Slowness

From an individual perspective, these slow cultural movements engender an ethical commitment to slowness. But for a planetary commons to come into sharper focus, it is a requirement to explore the collective and hence the social praxis of slowness. To revisit Elias and Moraru's articulation of the planetary from the introduction of this book, an ethical commitment requires the contagious spreading of the commoning process beyond our individual groupings, our communities, to make it as much a part of the functioning of the socio-economic world as possible. Hence, while these slow cultural pieces give us room as individuals to experience alternative worlds beyond capitalism's hyper-rapidity, they need to be applied to the 'everyday' realms of capitalist production.

Luckily, there are multiple movements around the globe that are already attempting this. For example, there is slow food, slow medicine and within academia there is slow scholarship.[13] I would like to focus, though, on slow fashion.

Slow fashion

In their annual report of 2018, the global exclusive fashion brand Burberry admitted to burning $36.8 million worth of stock, at the same time declaring revenue of $3.6 billion. After a boycotting campaign and the UK government voicing concern, the company announced that it would no longer burn its clothes. The reason why they burnt them instead of recycling or selling them on speaks to the wider issue of slowness. Essentially, burning stock so that no one else can have it is done to maintain brand exclusivity and therefore keep prices high. As an exclusive brand, Burberry's profit margins are based on artificial scarcity and the rapid change of styles. Fashion firms are beholden to the whims of the often unpredictable fluctuations of consumer taste, and that's not just the exclusive high-end brands. A few (like Burberry) may be able to set these trends, but the speed of fashion production has accelerated exponentially in line with the memeified nature of consumer culture, and so seemingly every fashion brand has to engage in 'fast fashion' to keep up.

In response to the increased need for cheap clothes produced quickly to satisfy the insatiable desire of the consumption class, the fashion industry is notorious for its use of sweatshops in the global north and south. When the UK city of Leicester went into local lockdown during the coronavirus pandemic in July 2020, it was believed to be because of the cramped and unsanitary conditions of the city's garment factories. Often employing people of colour (who are more at risk of contracting Covid-19 anyway) on low wages, the sector has over the years via a number of undercover journalist pieces been exposed as a crucible of fast-fashion-led exploitation.[14] So fast fashion is symptomatic of capitalism's contemporary consumption patterns; produce items quickly in response to the rapidly fluctuating (often online) market of aesthetic culture, use poor-quality but critically cheap material at a significant cost to the environment, manufacture the items using exploited labour, market them as disposable (again to the significant detriment of the environment), start the cycle over, rinse and repeat.

Hence, slow fashion was a movement that grew up around the resistance to this accelerated and unjust fashion production. The sustainability design scholar and activist Kate Fletcher has been writing about the slow fashion movement for over a decade and argues that slow fashion is not, and should not be, just fashion in reverse. It is a

Ethic 5: Slowness

'different worldview that names a coherent set of fashion activities that promotes variety and multiplicity of fashion production and consumption and that celebrates the pleasure and cultural significance of fashion within biophysical limits'.[15] Slow fashion, for Fletcher and for the wider global movement she has brought about, is therefore part of a broader set of questions that a slowness culture brings to the fore. It is not a trend to be plugged into existing production protocols of environmental damage and exploitation of poor labourers, it is to question the very nature and system of economic growth and the rapidity of consumer culture. Much like veganism, slow fashion, therefore, is a philosophical as well as a consumer movement. It aims to use locally sourced material, restrict 'new' lines to as infrequently as possible (often annual or biannual), operate as independents rather than chain stores and use fair wage structures. But these are just different consumption patterns that, while laudable, only offer a different way of consuming fashion, and are often far too expensive for most people. Fast fashion is popular because it is cheap; it offers people on low wages the ability to aspire to the cultural trends they are hawked on a continual basis by capitalism.

Slow fashion, therefore, is a structural movement to embed a slowing down of fashion consumption patterns across society. It is to systemically rewire the consumption of fashion that, like more and more parts of our consumption habits, has come to replicate the workings of the stock market. And here is the crux of the ethic of slowness: it rejects and goes against the constant fluctuations of the consumption landscape that are tied to the market. As Hayek's neoliberalism preached, and Thatcherism and Reaganomics actualized, our social world, including the fashions we buy into, have become marketized and fluctuate algorithmically in line with the multitude of variables that are generated by social media, the weather, presidential tweets, viruses and the constant circulation of memed information. The ethic of slowness calls for a stop to all this. A planetary commons, if it is to be a viable alternative to the capitalist world, *must* reject marketization and the rapidity of life it brings. Such a commons is to foster a collective dwelling in the present, with all the connections to the uncodified world it allows.

The power of the present

The Christian theologian Kathryn Tanner writes of how the present as a conceptual temporality has been completely redefined by capitalism, specifically by the form of financialized capitalism that neoliberalism has ushered in. In discussing traders on the stock market, she argues that they have to completely forgo any memories of the past or indeed thoughts of the future, and react within 'the edge of the about-to-be-past and the about-to-be-future'.[16] In other words, they have to forgo the connections that past successes or failures can have with present actions (just as throwing a 6 on a die 100 times in a row has no effect at all on whether the next throw will be a 6 or not). The market does not work predictably, so in order to maximize potential profits, traders must learn to segment themselves from the contiguity of the flow of social time and operate within the market's logic of time. Indeed, many financial institutions do away with the 'gut' feeling of traders altogether and have employed complex algorithms in high-frequency trading (HFT) techniques that can conduct millions of transactions a second. In this instance, the speed of decision-making is critical because milliseconds can make

Ethic 5: Slowness

a difference of millions of dollars. So much so that even the physical distance between the computers doing the HFT and the internet router is taken into account, given that the information will get to the more distant computers a fraction of a millisecond later. Even the speed of light, it seems, is too slow for financialized capitalism.

The financial markets – the crucibles of capitalism – do not require any longevity. Physical trading (retail, hospitality, the experience economy), for example, is subject to the whims of external factors that have longer-term cycles; for example, if it rains heavily one day, the high street footfall is down; if there's a global pandemic, then everything that requires physical consumption pretty much grinds to a halt. Yet the financial markets are not subject to these long-term externalities. They react to them, of course, but this is because they are predicated on volatility. As Tanner argues

> short-termism becomes not merely a way of avoiding risk but a primary means of profiting from the very volatility that produces risk. Prices go up and down, and one tries to capitalise on that very fact. Doing so, however, requires speed. Profiting from something in rapid movement requires equally rapid movement from you ... there is no profit in waiting.[17]

To resist this complete isolation of the present as a sacrifice to the scarcity of the market is to re-energize the present moment as *abundantly* connected to the past and the future. If the present, as a moment, is pregnant with the lessons of the past and the possibilities of a better future, then it is worth dwelling on the present to ensure those lessons are learned and those futures are realized as the best they possibly can be. In other words, being slow can enliven the abundance of the present to be put to work in a realization of a more just and equitable future, a planetary commons. With the wealth of energy that being radically connected to each other and the material world brings us as a collective, being slow actively can catalyse the provision of that energy. As Tanner goes on to argue, 'what prompts one to seize it right away is not the fear of missed opportunities, then, but the immediate, overwhelming attractiveness of the offer'.[18]

To recognize the abundance of the present is easy enough conceptually, but in practice it can be immensely challenging, both individually and collectively. This is because as much as capitalism has rewired us to act on short-term impulses, it has also very much commercialized our inevitable reactions to them. The wellness industry, linked to the happiness industry, offers us a chance to escape the stressful hypermobility of modern capitalist life, all for a fee of course. This industry has seen the appropriation of meditative techniques often derived from non-Western cultures such as indigenous peoples, Eastern philosophies and religions. Yoga, meditation classes, wellness retreats, digital detox programmes: there is a plethora of ways to de-stress and decompress from the constraints of working life, if only for a moment. They allow a 'recharge', a period of deep relaxation before heading back off into the throngs of capitalist realism. This clearly has relations to big pharmaceutical companies, with drugs and rehabilitation programmes being offered as a form of biopolitical and/or pharmacopornographic control.

But to disregard the benefits that these practices have for the self would be counter-intuitive. Of course they are helpful to us and have massive health benefits. They are critical survival techniques for many people on the margins of an unequal society, and sometimes they provide means by which anti-capitalisms can come to fruition. They offer

Ethic 5: Slowness

the chance to connect with distant and/or tacit atmospheres beyond our immediate perceptions and can offer different ways of existence that inculcate a radically different reading of the world around us. The depth of experience that well-being techniques can offer is articulated and narrated in different ways depending on our social and/or cultural condition – it can be prayer, meditation, Nirvana, grounding or even the 'runners' high' if running is your thing – but in all cases it connects us as individuals to the material and immaterial world around us. It can transcend the codified, quantified, individualistic, perceived world that capitalism has cocooned us in and offer glimpses of the infinite connections we have with each other and the earth as a living being, as Gaia. It can give headspace and emotional energy to the commoning practices needed to fuel a planetary commons.

However, the co-option of these well-being practices as another axiom of capitalist profiteering risks all this. Moreover, the increasing need for these techniques of mindfulness would not be there if the stresses of capitalist life were not so intense, unjust and unequal. Many workplaces and corporations will offer well-being programmes and meditation classes to build 'better resilience' way before recognizing that overwork and deadline pressures have caused those very stresses in the first instance. In universities, for example, the offer of petting zoos where students can come and 'de-stress' by close contact with aesthetically pleasing animals seems the current way to combat the mental ill-health epidemic. Not to downplay the benefit of cuddling a baby rabbit for a few minutes, but it is symptomatic of how institutions tackle the hugely problematic rise in mental ill-health: buy in external services from private providers that act as a sticking plaster, but continue to maintain the structures that create the need for them in the first place.

The ethical action of slowness reacts to this 'sticking plaster' mentality. So, for example, rather than petting zoos, thinking slow would be to address the relentless pressure put on students to perform academically (via the emphasis put on securing a well-paid job) and socially (via having to adhere to specific social-media-generated body images, fashions and trends). Being slow collectively, therefore, is more than simply reacting to the co-option of well-being practices by capitalism: it is to create new ways of organizing our institutions so that well-being doesn't have to be offered to negate stressful working practices. An ethic of slowness builds collectives and systems that foreground well-being practices for their commoning potentials rather than their resilience to an ever-more-unjust neoliberal capitalism. In the aftermath of Hurricane Katrina in 2005, the city's administration saw it as an opportunity to build a more 'resilient' city that could withstand the shocks of the economy and the natural world. The Black community, though, reacted to this with the slogan 'I am not resilient' because being resilient just means that something else can be done to them. The same is true of these well-being programmes; they should not be about building resilience yoked as it is to the neoliberal mantra of self-interest. Rejecting resilience and instead building a collective commoning practice of connecting the present to its multiple pasts and futures is what the ethics of slowness can allow.

Throughout this ethic, it has been highlighted how slowness is a critical check to the injustices of capitalism. Being slow is critical to realizing a planetary commons and the radically just futures it can engender. But like the paradox of capitalism itself, the ethic of slowness also can seem quite paradoxical given that in 2018, the Intergovernmental Panel on Climate Change (IPCC) gave the world only twelve years to reverse climate change before it would

Ethic 5: Slowness

be too late to avoid catastrophic damage. Given that we are already a number of years into that period, acting slow seems a luxury we cannot afford. However, *thinking* slow does not mean *acting* slow. Adhering to an ethic of slowness – dwelling in the here and now in order to experientially explore the connections beyond that – gives an impetus to change unjust capitalist structures even more quickly than we think we can. Demanding a slowing of production by changing our consumption patterns and actively tackling the injustices of 'fast' production can make a very rapid change to our world. The global 'lockdowns' during the coronavirus pandemic proved that we *can* slow right down and aid in the planet's recovery. During lockdown, CO_2 emissions plummeted, wildlife returned to the city, and we discovered new ways of feeding and providing for each other beyond capitalism. In the space of a few months, we had discovered a new way of being that highlighted the chronic injustices and inefficiencies of capitalism. Lockdown also caused job losses, mental health crises and thousands of needless deaths, so it is not to be fetishized. But the horrors of lockdown were a symptom of capitalism's inefficiency and the unwillingness of some of our more fascistic political leaders to change their destructive ways.

Slowness provides time for a planetary commons to breathe in the midst of what is a suffocating capitalist world. It also gives us the space to explore and test what works and, crucially, what doesn't. It gives us the space to fail. But failing is an ethic of its own to which we can now turn.

Ethic 6: Failure

Try again. Fail again. Fail better.[1]

In 2010, a neurobiochemist at the University of Edinburgh, Melanie Stefan, called on those in academia to publish their 'CV of Failure'. In 2017, she published her own and it outlined all the jobs that she applied for but didn't get, grant funding that she was unsuccessful with, academic papers that had been rejected and courses that she had failed to get into. Since then, the 'anti-résumé' has become relatively popular, with many supposedly 'successful' academics highlighting that they got to where they are nowhere near as straightforwardly as many will make out. What the CV of failure represents is that the 'path' to success is rarely linear, and involves multiple failures along the way. The moral of the story here is to not be too hard on yourself if you fail, to learn, and to do better next time. In a similar vein, within Silicon Valley, that most 'successful' of start-up hubs, since 2009 there has been an annual 'FailCon', a conference that brings technology entrepreneurs together to discuss their own failures, but also to 'prepare for success'.[2]

Publishing and discussing your failures from a position

Ethic 6: Failure

of success is no doubt an important practice for others in your field. It gives people starting out hope that it can be done, and also highlights that the sting of failure in whatever field – not getting your dream job, not getting the venture capitalist to invest in your latest app, getting that paper rejected, not getting as many views as you wanted on your latest online video – can be a learning curve if only you learn from what you think went wrong.

However, much of this rhetoric is couched in the already existing vernacular of what success is; a vernacular controlled by those in power. To be a 'success' (and therefore to recalibrate your failures to point towards that success) under neoliberal capitalism is yoked to the growth narrative. More money, more likes, more prestige: success is accumulation.[3] This vision of 'failing to succeed' is almost always accompanied with inspirational quotes and self-help narratives, the most common of which is the quote that opens this Ethic, from Samuel Beckett's novella *Worstward Ho*. The oft-quoted phrase has been used as a mantra in Silicon Valley and elsewhere as a call to use failure as a means to reconfigure your path to success. Failing 'better' is articulated as failing less short of your goal, and even less short next time, until eventually you succeed in whatever it is you – or perhaps more accurately capitalism – set out to do.

However, this capitalist reading of failure is folded into a single narrative: one that sees 'success' becoming ever more hoarded and accumulated. If Thomas Piketty's meticulous work has taught us anything it's that capitalism breeds sharpening inequality and a concentration of wealth in the hands of fewer and fewer people.[4] As the 'availability' of who can be successful therefore narrows, failure inevitably becomes more widespread. In my realm, academia, for example, there are now far more people vying for entry-level jobs than there were even five years ago, let alone when I was going for my first job post-PhD. The dual movement of universities attempting to recruit more postgraduate researchers because of the tuition they pay, and the lure of the academic life as a quasi-celebrity world, has meant there are far more people than there are jobs to fill. This inevitably leads to many post-PhD students leaving academia and therefore being perceived as a 'failure' by an academy steeped in neoliberal characteristics. The same process is at play across the economic world; capitalism catalyses and rewards the accumulation and the hoarding of wealth, but fewer people than ever get to share in it, and so this 'failure' spreads. No matter how many CVs of failure we read, it is unlikely to change this systemic injustice of the success/failure dialectic. In a finite world where the constant accumulation for some means intense violence for others, success can no longer be narrated this way.

The ethic of failure, then, is about rewiring success so that it no longer encapsulates the narratives of accumulation and growth. It is about celebrating and *actively* pursuing failure as a direct rejection of this kind of success. Seen ethically, failure is a counter-narrative to the growth-at-all-cost protocols of capitalism that have ushered in climate catastrophe. It is about realizing that there is already enough; there is no need for 'growth' that entails the further destruction of the planet and ourselves. Hence, the rest of this ethic will detail how to fail ethically. First it will articulate how failure is narrated under a capitalist discourse and adamantly linked to a self-interested individualism. It will go on to show how queering this success narrative opens up failure to be an ethical commitment to anti-capitalist thought. Finally, it will discuss the potentialities and pitfalls of how this can

Ethic 6: Failure

be 'scaled up' to think about failure as a collective, even at the level of the state.

Mind the gap

The concept of failure is often discussed within the literature as an individual trait, and yoked to a self-help discourse. There are countless 'gurus' out there willing to take your hard-earned income in order to tell you that if you're failing in life, really you're just on a different path to success from the one you envisioned. From Oprah Winfrey to Thomas Edison, there are countless rousing speeches, heartfelt emotional letters, anecdotes and vignettes about how someone has failed in what they wanted to do, only to find that in that process of failing, they managed to find other ways in which to succeed. And after they've quoted Samuel Beckett at you, they will perhaps use the next most famous quote about failure, this time from Thomas Edison, that goes something like, 'I have not failed, I've just found 10,000 ways that didn't work.'

This narrative of failure has acutely neoliberal characteristics. It pins 'failure' on the individual, and it ultimately refocuses the whole conceptualization of failure around improving the self, learning retrospectively and growing as a better – read 'more productive' – human being. The professor of art history Sara Lewis adheres to this mantra by arguing that failure is the gap between where we are and where we want to go, and only an 'individual knows if they are living in that gap'.[5] Living in this gap, she argues, helps us on the path to 'mastery', a phrase she says that we don't like to use, but one that is important if we want to succeed and grow. She talks of being uncomfortable, of having 'near wins', and uses examples from indigenous cultures and martial arts to highlight how failure or incompleteness helps us to achieve our goals.

There is some merit in these arguments. Despite this self-help narrative being completely embedded in the neoliberal drive for self-interest as the path to progress, it is important not to throw the baby out with the bathwater. Understanding our failures *is* an important part of how we develop and there is definitely an ethical dimension to learning how to cope with our failures. However, the response to understanding the 'how' of failure is to use that knowledge as further fuel to drive us on our already set path of self-fulfilment and individual growth, and thereby we ignore the 'why'. In other words, failing at a goal that we set ourselves should include questioning the goal as well. For failure, then, to be used as an ethic to enliven a planetary commons, it requires broadening out the narrative of failure to include not only introspection, but also a collective response; in a word, solidarity.

Solidarity is first and foremost collective struggle. Clearly it relates to the first ethic of mutualism, because to engage in solidarity in the first instance requires a letting go of the ideology of self-interest that has been thrust upon us by centuries of capitalist discourse. But beyond that, we often express solidarity at a time of failure. I distinctly remember that my first encounter with the word on a national level was in the aftermath of the 9/11 attacks in the US. I can recall seeing the front page of *The Sun* newspaper having a large US flag on it with the instruction to cut it out and hang it in your window to show solidarity with our transatlantic friends. On social media, we will often see political leaders and prominent commentators expressing solidarity with oppressed peoples in various parts of the world. But it is worth unpacking the term 'solidarity', largely because these uses of the word are 'performances' of solidarity on

Ethic 6: Failure

the curated stage of mass or social media. The indigenous Australian scholar Lilla Watson is credited with this quote: 'If you have come here to help me you are wasting your time, but if you have come because your liberation is bound up with mine, then let us work together.'[6] Solidarity, then, is more than offering help to people in need or at a time of their failure; it is an active struggle with someone or a group of people on whom your future depends. Flying the American flag in my window on 12 September 2001 would have been unlikely to be of much use to those who were fighting for their lives in the wake of the terrorist attacks. A UK politician who says on Twitter they stand in solidarity with various oppressed groups, such as Black female MPs or the LGBTQ+ community, when they are unlikely to suffer any of the constant tirade of abuse simply rings hollow.[7]

Solidarity, therefore, is about recognizing the vulnerability of the Other to hegemonic forces of oppression, and acknowledging your own complicity in that oppression, consciously or otherwise. It is to acknowledge your own failings as an ethical response to the injustices of capitalism wrought on others around you. Such an ethical failure was brought into sharp focus during the zenith of the Black Lives Matter campaign in the summer of 2020. Many white people expressed solidarity with Black people who were being killed by institutional police racism, yet stopped short of admitting their own complicity in a systemic whiteness that continues to oppress Black people across the world. As a white male academic myself, I wanted to express solidarity at what I saw as the horrific, unjust killing of Black people in the US. Yet without acknowledging that I benefit from a capitalist system that prioritizes white people when it comes to 'success' (i.e. jobs, wealth, criminal justice etc.), any such solidarity would simply be gestural. Therefore, to act in solidarity within this ethical framework of failure is to acknowledge that the system has failed Black people and to work to address this. It is to recast what success looks like within the system. To have personal wealth, promotion after promotion and research grant awards is not successful within a commoning discourse; what is successful is creating an academic environment in which Black people are no longer systematically and disproportionately rejected from jobs, senior positions and tenure. Failure, then, is not only acknowledging complicity, but rearticulating success along more just and equitable contours.

This redrawing of what success looks like therefore brings a planetary commons into view. Because the very nature of a commoning process looks to equity and justice rather than the constant search for growth and accumulation by dispossession, a 'successful' planetary commons will not be a simple case of more of the same. It is to embrace difference, and to not progress along a straight line of assumed success. A planetary commons will therefore 'queer' success.

Queering success

The 2006 film *Little Miss Sunshine* was the debut directorial film of wife-and-husband team Valerie Faris and Jonathan Dayton. The film tells the story of a bespectacled and tomboyish little girl, Olive, who lives with her family in Albuquerque, New Mexico. Her dad is a motivational speaker, although he fails to sell his 'success' formula and speaks to sparse crowds and polite applause. Olive's mum is an overworked housewife, and her brother has taken a vow of silence in order to achieve his goal of becoming a fighter pilot (he also reads a lot of Nietzsche). The family

Ethic 6: Failure

have also taken in Olive's uncle, a gay Proust scholar who has recently survived a suicide attempt, and Olive's grandfather, who has been evicted from his care home for snorting heroin. Olive dreams of competing in the Little Miss Sunshine beauty pageant in California, and so the extended family embark upon a road trip, in order to get Olive to the competition on time. As the journey progresses, each family member 'fails' in their own way: the father loses a potentially lucrative contract, the brother realizes he is colour blind and so will never be able to become a pilot, the uncle encounters his ex-partner who left him for his academic rival (prompting his suicide attempt) and the grandfather dies en route. Even the Volkswagen camper van in which they are travelling begins to break down. Olive appears to be the only one maintaining any hope of 'success'.

Once they get to the pageant, it is full of ultra-sexualized young girls, and it suddenly dawns on the family that Olive does not conform to the particular aesthetic of classic pre-teen all-American-ness and so will undoubtedly fail miserably. After the family attempt to talk her out of it, Olive gets up on stage anyway, and performs a subversive, but no less sexualized almost-burlesque performance. This horrifies the pageant organizers, who attempt to usher off stage. Olive's family take umbrage at this and join Olive to dance on stage, to the horror of the onlooking audience.

The film ends with the family accepting each other's failures, and their distinct marginality to the mainstream American vision of beauty and 'normality'. The family can be heard laughing with each other as their VW camper van careers through the parking garage barrier, unable to brake.

The film is a beautiful rendition of failure. Taking the heteronormative view of the traditional nuclear family, it slowly unravels it and uses the revelatory-road-trip film genre to develop a realization in each character of their role in the family and in life more broadly. The film narrates how as individuals they have failed in their goals, but as a collective family, they utilize that failure as deviance from the mainstream cultural hegemony of a patriarchal sexualizing America. In many ways, the film uses failure as a mechanism to articulate how it can transcend and *queer* dominant narratives of what success is within capitalism.

The queer theorist Judith (also known as Jack) Halberstam (who also uses the film to explain her theory in a far more eloquent way than I do) has argued in her book *The Queer Art of Failure*: 'As a practice, failure recognizes that alternatives are embedded already in the dominant and that power is never total or consistent; indeed failure can exploit the unpredictability of ideology and its indeterminate qualities.'[8] This is a view shared by cultural theorist José Muñoz, who argued failure is the 'always already' of minoritarianism within dominant culture.[9] Failure, then, can be seen as 'simply' non-compliance with the dominant order, being resistant, irksome, passive or inactive – what the political theorist Alan C. Scott famously called 'weapons of the weak'.[10]

Within *Little Miss Sunshine*, the 'already embedded' nature of each character's failure slowly reveals itself first as a horrifying incapability to achieve what capitalist life has dictated they must. But then that 'failure' coalesces collectively to create a deviance that they accept, embrace and live. What is more, they realize this by exploiting the ultra-sexualized pageant with Olive's own brand of sexuality, one that is deviant and anti-feminine (for the dance's opening Olive wears a tie, a waistcoat and a top hat). They disrupt the 'smooth' functioning of the pageant by dancing

Ethic 6: Failure

on stage together in an unchoreographed, improvised mess of flying ecstatic limbs. Indeed, as Halberstam writes about this iconic scene, this failure is 'so much better, so much more liberating than any success that could possibly be achieved in the context of a teen beauty contest'.[11]

Halberstam couches this analysis of the film in the broader argument that queerness is a deliberate search for darkness instead of light and the perceived inadequacies it can highlight, moving away from straightness and the conformity that comes with it. She argues that 'queer art has made failure its centrepiece and has cast queerness as the dark landscape of confusion, loneliness, alienation, impossibility, and awkwardness'.[12] *Little Miss Sunshine* exemplifies this queer art in so far as it focuses on the awkwardness of the characters and their alienation from mainstream life. In other words, they are a family of misfits, but that is totally okay.

So within this ethic, failure is viewed as counterpoised to the success-drive of capitalism. As Muñoz and Halberstam argue, it queers the straightness of the mainstream, and creates deviance that gives space for subculture and minoritarianism to gestate, infect and slowly spread into the mechanics of the mainstream. It is not simply recasting success as a different end goal from yours, such as having a different political party in charge, a different team winning the league, or even having a Black female as CEO instead of a white man, however progressive that success looks. It is about using failure as a way of questioning whether we need political parties, leagues or CEOs in the first place; or at the very least questioning their rules of operation and engagement. Failure, then, when wrenched out of the vernacular of capitalism, becomes fertile ground for revolutionary thought.

Seven Ethics Against Capitalism

Failure loves company

It is therefore no wonder that many scholars will speak of supposedly 'failed' revolutions as nothing of the sort. They become educational moments in the ongoing struggle against an ever-changing and predatory capitalist hegemony. When writing about the Paris Commune, the professor of comparative literature Kristin Ross argues that it has too often been historicized as a heroic failure. Often couched between the twin narratives of failed state socialism and a radical moment in the establishment of a French Republic, the Commune is cast as a lesson for future socialist movements (or as a warning for future revolutions), but a failure nonetheless. It is this narrative that Ross tries to get away from when researching the Commune and the writings, political imaginaries and practices of the Communards who populated it.

In her book *Communal Luxury*, which details the lives and legacies of the Communards, the 'failures' of the Commune are re-narrated as seeds of revolutionary thought that have catalysed many of the anti-capitalist movements that we still see today. As the introduction to this book highlighted, the Paris Commune is often thought of as part of the history of the commons, but this can only be a productive exercise if the narrative of failure (from both a capitalist and a socialist state reading) is meticulously extricated from it. For example, Ross argues that the commoning practices that the Commune enacted blurred the boundaries between hitherto immutable institutions such as education, art and politics. As she argues;

> More important than any laws the Communards were able to enact was simply the way in which their daily work-

Ethic 6: Failure

ings inverted entrenched hierarchies and divisions – first and foremost among these the division between manual or artistic and intellectual labour. . . . What matters more than any images conveyed, laws passed, or institutions founded are the capacities set in motion.[13]

Dwelling on the Commune's 'Manifesto of the Federation of Artists', Ross states that the Communards were using the social experimentation of the Commune to completely 'deprivatize' art and beauty and were integrating them into everyday and educational life. The manifesto was an attempt by the artists of the Commune to therefore proletarianize artistic production, to take museums and monuments back into public and working-class ownership, and have the art and architecture on display subject to democratic control.[14] This reconnection between art, the public and the common sphere was an important moment of the Commune, and one that is vital to the realization of a planetary commons more broadly. Not wanting to revisit vast, voluminous work that articulates the role of creativity from an artistic and social perspective as critical in realizing alternative forms of the future,[15] it is sufficient to say that such a view was partly forged in the revolutionary fires of the Paris Commune. But as the cultural theorist Max Haiven has argued, the event has been historically rewritten as a violent outburst on the way to the capitalist End of History, a simple building block that only fits into the grand narrative of the present immutable status quo in one way.[16]

But recasting the Commune away from the simple dialectic of a failed communist utopia or a corrective to capitalist realism, towards being a historical event latent with revolutionary energies that can power emancipatory thought now and in the future, is vital. Indeed, more recent histories have seen the Commune invigorated, reignited and reincarnated in various forms all over the world. The fact that it is still heralded as a monumental point in revolutionary history speaks to its transformative power. In the wake of the Black Lives Matter riots of 2020, a number of 'autonomous zones' were set up in North American cities that deliberately emulated the politics of the Paris Commune to create enclaves of anti-racist discussion and celebration of Black culture.[17] Another, older example is Christiania in Copenhagen, which lives on as one the most famous counter-cultural and subversive communes in Europe. In 1972, a group of artists, anarchists and radical revolutionaries from across Denmark squatted an old army barracks in inner-city Copenhagen. In the near half a century since, it has struggled (largely peacefully) with the city authorities and remained a site of commoning, where private land has been abolished, mutual aid is the driving force of provision and soft drugs can be consumed 'legally' without threat of arrest. The debates about the ongoing gentrification of the site have been long-standing, not least in 2011, when those living at the site lost a court case for it to remain unprivatized. The land therefore became 'owned' and hence managed by a committee made up of members and outsiders. Since then, it has slowly begun to change how it operates; buildings require official regulations, national bylaws and criminality can be applied, and the site now has to raise funds to maintain its functioning as a democratic entity.[18]

Despite the legal wrangling behind the scenes, the site still very much adheres to the aesthetics of its original set-up; it is a site where misfits, counter-cultural and subcultural groups, revolutionaries, anarchists and anti-capitalists can meet, debate and live in (relative) freedom from the pressures of the urban capitalist gentrifying system. Compare it to other squatted sites across Europe

Ethic 6: Failure

such as Grow Heathrow (discussed in Ethic 2), and we can see it has managed to stave off the threat of eviction and displacement relatively 'successfully'. To do so, however, it has had to take up some elements of the very system it was attempting to extricate itself from, namely private property and city governance structures, leading to accusations that it has 'failed' to maintain its subversive ethos and embraced neoliberal forms of governmentality.[19]

Viewed through the lens of this ethic of failure as part of the commoning process of constantly searching for equality and justice, Christiania can be viewed as beyond this dichotomous success/failure story altogether. It has forged a new urban political imaginary that problematizes the smooth functioning of capitalist urban governance, yet at the same time has also attempted to infuse its subversive and counter-cultural ethos into an institutional setting. In analysing Elinor Ostrom's discussion of the commons (which I outlined in the introduction), the Marxist geographer David Harvey has outlined the 'scale problem', in that 'what looks like a good way to resolve problems at one scale does not hold at another scale'.[20] With Christiania, we see this in practice in so far as the 'feel' of its commoning and anarchist roots have been lost with its 'progression' into an institutionalized form of private ownership. No matter how democratic the process is of governance within the commune, it is inescapable that they have yielded to the forces of property ownership in order to safeguard a particular way of life.

A similar process has taken place with a site dear to me, the skate spot on London's South Bank (which I have written about in the past[21]). Since it was 'saved' back in 2015, the site has 'grown' into an official part of the South Bank's leisure provision. The site has been expanded to its original specifications and size, but additional health and safety measures have been put in place, there are programmes now on offer, and generally it feels like a far more 'official' site than it was previous to the 'battle of the South Bank'. In these cases, and many more besides, the dualistic narrative of success versus failure has closed off the commoning practices. This is because the 'institutionalization' of these sites has allowed the equity, justice and ethical forbearance that an anti-capitalist viewpoint brings into focus to percolate into the previously highly gentrifying and capitalist governance networks and structures. But these will sit alongside the obvious problems of gentrification, hyper-tourism and police surveillance, so this is not to say there has been no 'dilution' of the progressive ideals; it is always a battle to maintain a fidelity to the event that created the site of commoning in the first place. But even in these small locales, there are tangible examples of the ethics of a commoning practice – some of which look similar to those already described in this book – seeping into otherwise rigid and *un*ethical capitalist structures. The city government of Copenhagen, for example, is actively pursuing the legalization of marijuana in response to (and perhaps to muscle in on) the illegal, but tolerated, trade of the soft drug in Christiania. The South Bank has started to include skateboarding programmes as part of its youth-engagement and outreach activities, thereby attracting children who would otherwise not engage with a large cultural institution.

To couch these sites in terms of whether they have 'succeeded' or not is to miss the nuances of the commoning practices they have engendered, and where they need more work. The tireless work of activists, squatters, campaigners and commoners in setting up and initiating sites of anti-capitalist practice (often at great physical and psychological cost) is an energy that is vital in commoning. It is the vital energy needed to escape through the thick layer of

Ethic 6: Failure

capitalist realism that pervades space; without these commoning pioneers (like the Diggers and the Communards in the past, and the Christiania squatters and the protesting skateboarders in the present), capitalism would be even more pervasive than it already is. But these fugitive currents of energy can fizzle out if there isn't an institutional 'safe space' that is at least willing to recharge them. So we need, in essence, a 'failed' state.

The failed state

To put it another way, can this articulation of an ethical form of failure 'work' at an even larger scale, perhaps even the scale of the state? Combining the term 'failure' with 'state' is a geopolitical ploy, one that conjures images of despotic governments, often in the global south, mass poverty, failing infrastructures and civil unrest. There is even a codified index created by the OECD (Organisation for Economic Co-operation and Development) that classifies countries around the world as to their 'fragility', based on indices such as the flow of finances to and from countries and how 'stable' they are (i.e. the level of corruption), civic unrest, inequality and environmental damage. It is telling that during the coronavirus pandemic, the Black Lives Matter uprisings and the pro-Trump invasion of the Capitol, the violent and incompetent response by the then-incumbent Trump administration has seen many mainstream outlets labelling the US as a 'failed state', given that it scores highly on the very same indices it uses against supposedly fragile states in the Middle East, Africa and other parts of the world it polices.

So within the global view of capitalist expansion, there is seemingly no way in which a country can be allowed to fail in the ethical way described thus far. Any country that attempted to replace the growth narrative of ever-expanding GDP (gross domestic product) with one of replacing capitalism altogether would automatically be castigated as communist, revolutionary, backward and in immediate need of invasion and conversion to a capitalist way of life.[22] This has not stopped countries such as New Zealand pursuing a 'degrowth' agenda that puts citizens' health and well-being before GDP growth as national budgeting goals.[23] And, in the chaotic beginnings of the coronavirus pandemic in the spring of 2020, Dr Mike Ryan, the Director of the World Health Organization's Health Emergencies Programme, said this in one of the WHO's daily press briefings:

> One of the great things in emergency response ... if you need to be right before you move, you will never win. Perfection is the enemy of the good when it comes to emergency management. Speed trumps perfection. And the problem we have in society at the moment is that everyone is afraid of making a mistake. Everyone is afraid of the consequence of error. But the greatest error is not to move. The greatest error is to be paralysed by the fear of failure.

Those countries that 'successfully' dealt with the virus in terms of minimizing deaths and economic impact were those that acted quickly, and were not paralysed by the fear of failure. It was telling that countries that are bastions of the most populist versions of neoliberal capitalism, the US and UK, were badly affected because they were seemingly too slow (or unwilling) to act. Within the setting of a global pandemic in which the smooth functioning of capitalist life actually helps to spread the virus and makes matters worse, failing was an important feature of 'defeating' the virus. To save lives, states needed to not be scared

Ethic 6: Failure

of failing *ethically*. And as the post-virus world comes into focus, where the ideologies of capitalism and unfettered growth will become ever more the object of political, social and cultural scrutiny as we hurtle towards climate catastrophe, ethical failure at the state level will become vital.

As such, a planetary commons – in which failing ethically is an important energizing factor – can be thought of as a real political alternative to a seeming capitalist realism. If states can implement failure as part of their apparatus in a pandemic, then they can do so in response to climate change. And such 'failure' will necessitate the questioning of the voracity of capitalism. In the process, however, states will need the ethical commitment of those locals and other individuals who have been using all their energies in fighting for a just and equitable world-in-common all along.

As Halberstam writes, 'there is something powerful in being wrong, in losing, in failing, and that all our failures combined might just be enough, if we practice them well, to bring down the winner ... Failure loves company.'[24] And so a state-led ethical failure can energize the commoning practices of its citizens. Of course, there is a danger that people and institutions will try to convert commoning practices into capitalist energies (via capitalist appropriation and co-option, or simply by selling out), but without that release of commoning energy into the 'wider' world, large-scale change will never come. The process of collective commoning is not without risks, one of which is indeed the risk of 'failure' to topple the winner that is climate-catastrophe-inducing capitalism. But to return to Beckett's quote that opened this chapter, with hopefully a different, more ethical meaning, 'Try again. Fail again. Fail better.'

Ethic 7: Love

Below is an extract from The Message translation of the Bible, of 1 Corinthians 13, and for reasons that will become apparent as this Ethic goes on, it is worth quoting in full.

> If I speak with human eloquence and angelic ecstasy but don't love, I'm nothing but the creaking of a rusty gate. If I speak God's Word with power, revealing all his mysteries and making everything plain as day, and if I have faith that says to a mountain, 'Jump,' and it jumps, but I don't love, I'm nothing. If I give everything I own to the poor and even go to the stake to be burned as a martyr, but I don't love, I've gotten nowhere. So, no matter what I say, what I believe, and what I do, I'm bankrupt without love.
>
> Love never gives up.
> Love cares more for others than for self.
> Love doesn't want what it doesn't have.
> Love doesn't strut,
> Doesn't have a swelled head,
> Doesn't force itself on others,
> Isn't always 'me first',
> Doesn't fly off the handle,
> Doesn't keep score of the sins of others,
> Doesn't revel when others grovel,

Ethic 7: Love

> Takes pleasure in the flowering of truth,
> Puts up with anything,
> Trusts God always,
> Always looks for the best,
> Never looks back,
> But keeps going to the end.

> Love never dies. Inspired speech will be over some day; praying in tongues will end; understanding will reach its limit. We know only a portion of the truth, and what we say about God is always incomplete. But when the Complete arrives, our incompletes will be cancelled. When I was an infant at my mother's breast, I gurgled and cooed like any infant. When I grew up, I left those infant ways for good.

> We don't yet see things clearly. We're squinting in a fog, peering through a mist. But it won't be long before the weather clears and the sun shines bright! We'll see it all then, see it all as clearly as God sees us, knowing him directly just as he knows us! But for right now, until that completeness, we have three things to do to lead us toward that consummation: Trust steadily in God, hope unswervingly, love extravagantly. And the best of the three is love.[1]

The ancient Greeks had four words for 'love': *eros*, *philia*, *storge* and *agapē*; each one implying a quite radically different meaning of that most visceral and esoteric of human emotions, love.

The first, *eros*, is perhaps best known to Anglophone speakers. Eros was the Greek god of love (the Roman equivalent being Cupid), and has been depicted throughout antiquity as a winged cherub. But the word *eros* stood for a passionate, sexualized love (hence the term 'erotic'). This is not only a love that one individual has for another, driven by biological and physiological desires, but also a compassionate love of another individual, a long-term partner. In both cases, though, it is an individualized love, devoid of any connectivity external to that desire of self-sustenance.

This kind of love is a politically and socially indifferent force, but only fairly recently so. The 'free love' movement that accompanied the hippie subculture in the 1960s was seen as revolutionary. It rejected societal technologies of love, such as marriage, commitment, family life and the rearing of children, and monogamous relationships. It was a reaction to the perceived institutional, state-led and class-based shackles that directed sexual acts to being part of a familial reproductive programme, something to produce more workers for the capitalist system.

However, with the rise of postmodern capitalism and the appropriation of critical movements, such sexual freedom was reduced to a commodified desire. The Croatian philosopher Srećko Horvat has succinctly argued that the May '68 uprisings in Paris, based as they were on a desire for freer love between students, was a time in which the force of *eros* could have been a revolutionary force for real change in society.[2] However, as we now know, this period saw the complete commodification of love by the new, agile, flexible forms of capitalist appropriation. Now, in the digitally mediated age of the twenty-first century, the radicality of free love has been appropriated by a notion of promiscuity. Sexual freedom has lost its revolutionary capacities, and instead become absorbed into the capitalist dialectic. On the one hand, promiscuity, particularly in women, is utilized as an accusatory gesture by mainstream conservative values. It has even been used to excuse rape and physical abuse. On the other hand, the onset of dating websites, and location-based dating apps such as Tinder and Grindr, have commodified the encounter, taken away

Ethic 7: Love

the 'fall' into love, and replaced it with what Horvat has called the 'swipe' into love. So the erotic love that the Greeks spoke of has been an attribute of capitalist accumulation and its masculine bias for many decades.

The second version of love that the Greeks articulated was *philia*, meaning friendship. We see this mostly as a suffix to denote a love for something in particular; for example, a Francophile being someone who loves all things French, or technophilia, which is the love of technology. The Greeks, though, originally used the term to express a deep friendship, a 'sisterly or brotherly love' (an etymology that gives us the 'city of brotherly love', otherwise known as Philadelphia). Such a love was more than familial, however, as it was free from 'natural' ties. You can't choose your family, but you can your friends. In this sense, *philia* expresses a companionship devoid of erotic love; a love that is in many ways 'unnatural' (in that we don't need friends to reproduce as a species). Such a love is appreciative, reciprocal, productive, even mutual, and is utilized to achieve a shared goal; in many ways it is solidarity; a political, class-based love. As discussed in the previous Ethic of failure, solidarity is a vital concept of the commoning process because it rallies others to the revolutionary cause. Hence there is much to be learned from this reading of a *philia* articulation of love.

But the term *philia* has been used with opposing meanings at times. The cultural critic Edward Said has alluded to a similar ideal as 'filiation', which he cites as more of a naturalized form of kinship (perhaps kinship with a sister or brother), in contrast to 'affiliation', which is something constructed socially and transpersonal, often to replace 'filiation' (for example, friends from school or work with whom you form a social network that benefits all of you in some way).[3] Yet, in contrast to this, the philosopher

Jacques Derrida's famous essay *The Politics of Friendship* declares that friendship (or more accurately, fraternity) is a critical component of a more radical form of politics and democracy.[4] To think fraternity beyond biological ties is to acknowledge someone as equal, another self. Derrida invokes Aristotle's notion of friendship as a fundamental part of democracy. Indeed, a shared citizenship is only possible once a 'canonical' friendship is forged, one that invokes a shared origin that has 'constancy beyond discourses' such as biological fraternity (i.e. beyond discourses that include being 'naturalized' friends via similar biological characteristics, such as having the same skin colour). Derrida stresses that a *philia* or filiation that is not limited by biological discourses is able to maintain a critique of friend/enemy logic, and thereby maintain the possibility of transcending that binary and moving beyond wars and enemies. So fraternity extends beyond bloodlines and shared DNA; it is a space of possibility. It is little wonder that the slogan of the French Revolution was 'liberté, égalité, *fraternité*'.

However, like the free love movement, such a fraternity that transcends biological links and seeks a common interest can have destructive powers when utilized intensively. One of the critiques of neoliberal capitalism is that it is highly nepotistic. It thrives on the imposition of 'cliques' in order to hoard power rather than to fight it. This is the distinct opposite of solidarity. This is perhaps most readily recognizable in the accusation that political elites are often chosen from existing elitist networks. For example, nine of the last twelve British prime ministers studied at Oxford University. And one only has to take a quick glance at the people Donald Trump hired (and then very quickly fired) in prominent positions in his chaotic administration after he entered the White House in 2016. Rex Tillerson,

Ethic 7: Love

the ex-CEO of ExxonMobil, became Secretary of State; Steven Mnuchin, a former banker for Goldman Sachs, became Secretary of the Treasury; and Wilbur Ross, a former multimillion-dollar investor, became Secretary of Commerce. These were all people with whom Trump had done business in the past. Indeed, Trump is only one in a long line of American presidents to fill his administration with personal contacts, and there are many different examples that could be used (both Republican and Democrat). In a more direct and striking example, in 2013, the UK's national postal service, the Royal Mail, was floated on the stock exchange as a public company. The then-chancellor, George Osborne (who, incidentally, was also educated at Oxford University), was accused by many in the press of pushing the sell-off, as it resulted in a huge financial windfall for his best man, Peter Davies. Davies' hedge fund company, Lansdowne Partners, made £36 million from the sell-off. And for sure, there will be countless more examples like this that remain uncovered and unrecorded.

Suffice to say the term 'revolving door' is used liberally to imply the reciprocal networks of power that span government, lobbyists, business and senior figures in public institutions, and allow for wealth and benefit to flow at the expense of those outside them. These cliques are a prime example of how the political will of (a particular narrow definition of) fraternities can be utilized to centralize resources, distort democracy and perpetuate injustice. Furthermore, there is a gendered bias to all of this; 'fraternity' is of course a masculine term, and the corridors of power in many of the world's national governments and global corporations are overwhelmingly populated by white men keen to exclude anyone who challenges that power imbalance.

Related to *philia*, but not entirely the same, the third

definition of love that the Greeks articulated was that of *storge*, loosely meaning a familial love, specifically between a parent and a child. It refers to a love that is abundant, but brought about 'naturally'. A parent's love for their child is as much a product of biological and neurochemical responses as it is of familiarity. Of course, this need not always be via bloodlines, as foster parents, adopted guardians and inter-generational carers also develop *storge* in response to prolonged intimacy that replicates the more normalized nuclear family setting. Invariably, the socio-cultural setting of the family will directly influence the practice of *storge*, with the variance in cultural and national 'versions' of parental behaviour an obvious barometer (one only has to think of the stereotypes of different nationalities depicted in popular culture).

For the mobilization of neoliberalism that fuelled the growth of capitalist accumulation from the 1980s onwards, the notion of the family has been critical. Thatcherism, described in the first Ethic of mutualism, was predicated on a reliance on the family unit. This classic trope of conservatism pervades neoliberal ideology, with many state programmes at the time (and subsequently) basing welfare payments on the traditional gendered family unit. Within the Anglo-American form of neoliberalism from Thatcher and Reaganism onwards, the family (for both blue and red political hues) has formed the basis of the expansion of neoliberal policies. From tax breaks for married couples, to forcing single mothers to interact with the biological father of their child(ren) whether they want to or not in order to get support payments, the state has always been a factor in enforcing the family unit. Why? The neoliberal apologists will tell you it is because the family unit instils responsibility, discipline and a moral compass that creates socially and economically valuable individuals. However,

Ethic 7: Love

as the political scientist Melinda Cooper has argued, the family unit in capitalism is the site of the 'externalities' of free-market trade, namely debt, care and the transfer of wealth. Often forged from a very *un*holy alliance with conservative religious factions (notably in the US), capitalism anchors the family as the mechanism of social reproduction. It is the responsibility of parents to care for their children, to put them through school and higher education – and, crucially, to pay for it. Cooper focuses on the role of student debt, which 'is increasingly a family affair, keeping parents, children and relatives enmeshed in webs of economic obligation for decades on end'.[5] Given that the state-funded student uprisings of the 1960s and 1970s across the world were resolutely anti-establishment, an obvious counter-attack would be to put the burden of the cost of education on the immediate family. This (along with the broader marketization of higher education across the world) has had the rather useful side effect of creating a generation of students who are more focused on career prospects than creating a better, anti-capitalist world.

The family unit, as many feminist and queer scholars have discussed, is a fundamentally capitalist institution because it is completely structured to create productive workers who maintain the property-owning, resource-extracting and privatized social fabric that ensnares everyone in the capitalist trajectory. Indeed, feminist activist Sophie Anne Lewis has argued for 'family abolition', which is the 'end of the double-edge coercion whereby the babies we gestate are ours and ours alone, to guard, invest in, and prioritise'.[6] And as Thomas Piketty's work has shown, inherited wealth, most often via familial relations, is the main source of inequality in today's world. All this is not to deny the *storge* form of love that can exist between family members; it is very *real*. But such an understanding of love in this way can ossify the structures of capitalist growth, rather than challenge them.

Other articulations of *storge* pertain to it as the 'friendship' love; a love between two people that starts as friendship but then develops into a deeper, perhaps even romantic love (and then back again). Often devoid of sexual intimacy yet nevertheless practised between the traditional 'couple', such a love is dependent upon the passing of time. It is often characterized by couples that remain good friends even after the 'break-up' of their romantic or erotic relationship. But like the erotic versions of love, this version of *storge* has also become another axiom of consumerist-soaked late capitalism. Friendships that are deep, detailed, discursive and lovingly curated over time are being replaced by a more nebulous and vacuous notion of virtualized friendship. Facebook is the main culprit, of course, with its very vernacular of 'friends'; the development of our friendship circles becomes nothing more than an exercise in data gathering and a constant performance of what an idealized friendship should entail. Other social media and instant group-communication platforms have come to dominate the interface of friendship and in so doing stripped communication of the visceral, non-verbal, affective, place-based encounter. 'Friendship' can now be measured in how quickly you respond to each other on WhatsApp or how you react to a 'friend's' post on Facebook. And despite these being called 'social' media, loneliness and mental health epidemics, particularly in the young, are at record levels.

Then there is *agapē*, which, etymologically at least, pertains to the love of humankind for God, and God for humankind. It is the kind of love that we read about in that famous Biblical passage from 1 Corinthians (that opens this chapter), which no doubt many of us would

Ethic 7: Love

have heard at a Christian wedding ceremony, if we've ever attended any. The irony is, of course, that the love talked about in this passage has very little to do with the love that is forged between two people in marriage, which is more akin to the *eros* version of love described above. The *agapē* described by the Greeks and in the letter by Saint Paul to the church in Corinth is very different indeed.

And such a reading of love need not be limited to religious discourses either; it is a love that is not so easily co-opted by soft hegemonic powers because it denotes an *unconditional* love. *Agapē* thrives when something is given without any semblance of reward; hence it is completely antithetical to contemporary socio-cultural renderings of a conditional or self-sustaining love. A love that is selfless and ubiquitously equitable is the exact opposite of late capitalism, which thrives on selfishness, conditionality and inequality. It is a love that has no conditions, and as such has a radically generative power to create things that are entirely new. It is, as Horvat so beautifully articulates it, a revolutionary love.[7]

Such a love has a clear resonance with Christian-taught Gospel teaching. Yet certain factions of the institutionalized Christian religion often preach nothing of the sort They have been responsible for numerous bloody Crusades, been used to justify slavery, turned a blind eye to historic child abuse claims, and in their most extreme forms have mutated into fascist tendencies in Nazi Germany and Trump's America. *Agapē* love that incorporates the Gospel teachings of Jesus is antithetical to organized religion and indeed any form of hierarchy that subordinates one group of people to another. It is distinctly anarchist in its practice; it is a form of christianarchy.

Jacques Ellul, a prominent christianarchist, argued vociferously for a love that was more than generous giving, sympathy or pity. His argument was for a radical love forgoing solid, comforting ground for the uncertainty of poverty, marginality and unstable places. Ellul brought into collision the teachings of Jesus and Marx, and argued that the love shown in the Biblical Gospels is a subversive force, one that rejects the very human articulations of morality, law or ethics. Indeed, he argues: 'Love, which cannot be regulated, categorized, or analyzed into principles or commandments, takes the place of law. The relationship with others is not one of duty but of love.'[8] So Ellul was following a lineage of radical Marxist and/or anarchist thinkers (notably Leo Tolstoy) who used the radicality of Jesus' teachings on love as a means to articulate it as a subversive practice rather than an emotion, or affective register. Ellul was heavily involved in the French resistance to Nazi occupation during World War II, sheltering Jews and providing them with false papers to escape encampment. He also set up clubs for juvenile delinquents, resisted large-scale developments that would cause environmental damage, and was supportive (by means of his training as a lawyer) to conscientious objectors, on the basis of his commitment to nonviolence as a form of civil disobedience.[9]

A similar kind of love can be found in the words and actions of the civil rights activist Martin Luther King Jr. He is most famed for his 'I have a dream' speech, delivered at the March on Washington for Jobs and Freedom in August 1963, but his philosophy on love is equally radical and engaging. Four years after that famous speech, a few months before he was assassinated, he said: 'Power without love is reckless and abusive, and love without power is sentimental and anaemic. Power at its best is love implementing the demands of justice, and justice at its best is power correcting everything that stands against love.'[10] King was critiquing Nietzsche's rejection of

Ethic 7: Love

Christian interpretations of love as overtly nihilistic, as well as the theological position of many Christians who rejected Nietzsche's arguments that life is 'simply the will to power'.[11] As such, King was combining the two concepts to suggest that only then does the power of love overcome the love of power. King's philosophy of the power of love builds upon the work of the theologian Paul Tillich, who defines power as simply the drive of everything living, both human and nonhuman, to 'realise itself with increasing intensity and extensity'[12] and love as a 'drive towards the reunion of the separated'.[13] Crucially, union does not mean sameness, but a union of difference and the entirely new creative subjectivities that it can engender. Tillich, himself discussing *agapē*'s version of love, brings together the two concepts of power and love under the rubric of justice, which is power being realized and actualized by all living things with love, that is, to be united. Justice is therefore the drive for all beings and things to exert their power to unite together and create new worlds outside of the existing hegemony of the status quo (in our case, capitalism). Justice can hence be violated any time that one form of power overcomes or denies the power of another.

Moreover, both King and Tillich have argued that this love cannot be a 'self-love', because if the power of love is used to raise up the self, it is denying unification. 'Self-actualization' and 'self-care' would be better articulations of this, and indeed have their place in the act of commoning. It is important to promote self-care as it can be the vital rest and resource that are required when people are in danger of burn-out. Activism is fraught with tireless campaigners suffering undue mental ill-health or physical illness that can only be rectified by a period of self-care. Withdrawing from the struggle momentarily, taking time to recharge with an ethical slowness, drawing on the energies of the materials around them are essential for the commons to ethically infect the capitalist world around them. As Ethic 1 outlined, being mutual is tiring work. And so self-care is a vital aid for those people looking to engage in commoning practices long-term (as long as it is practised without falling prey to the capitalist narrations of 'well-being', as Ethic 5 outlined). But it is not the kind of love that will propel the planetary commons; only an a*gapē* love can achieve this.

This is because an *agapē* reading of love is a powerful rendition that is completely selfless, based in praxes of solidarity, and compels us to forgo comfort in order to help others and search for justice that comes from union (as Tillich outlined). It is in complete opposition to how society is structured today. As Ethic 1 described, neoliberal versions of capitalism are founded on self-interest, but thinking of how to decentre the self and be mutual requires an agapetic love that engages in making ourselves 'uncomfortable' (if only temporarily) so as to connect with others and to reunify the separated. As we saw in Ethic 2, acknowledging how materiality affects and energizes us is fuelled by the praxis of *agapē*. With the third ethic, connecting with minor subjects and those oppressed and marginalized systematically requires *agapē*, as without it, we would only see like-minded people and exist in our own comfortable echo chambers. As Ethic 4 highlighted, moving away from a codified self and a quantified society will necessitate a submission to *agapē* and its complete inability to be regulated, codified, quantified or categorized. With Ethic 5, being slow requires patience and a willingness to simply 'be' with yourself and others, which replenishes *agapē* and recharges commoning practices. And as Ethic 6 showed us, failing, but failing without succumbing to the meaning of it as simply another version of

Ethic 7: Love

capitalist success, requires the collective energies that only an agapetic form of love can bring.

Agapē, then, is marbled through the rest of the ethical commitments outlined in this book. It is the energy that fuels commoning practices and keeps a planetary commons – with its commitment to the community and a common resource dialectic – firmly on the horizon. But it can only be fuelled itself by further ethical praxes. Hence, it is active in resisting capitalism totally. By forgoing the self in favour of the Other, by focusing on people that do not conform, by dwelling in the here and now, by extending to nonhuman matter, by making time and room for failure, and, more generally, by resisting the line that we must constantly 'grow' and 'accrue' and instead forcing a deeply intimate, socialized and decodified relationship, *agapē* is antithetical to contemporary models of social organization under capitalism. It generates newness in the face of a homogenizing hegemony. As Hannah Arendt showed us via the concept of natality[14] (which is the concept of 'being born as a human' as the beginnings of our capacity for newness), being born into the world is the beginning of our ability to create something entirely new, even a planetary commons. So if being born is the genesis of all newness in the world, it is no wonder that it is the result of *making love*.

Examples of this *agapē* exist all around us, within the fissures of neoliberal capitalism and in the gaps between the observed parts of our world. From the Black Lives Matter movements, thousands upon thousands of photos and videos circulated on social media that depicted the protests and the riots. Most were of horrific acts of police brutality, but others were genuine acts of love. One in particular that stayed with me was of a Black, seemingly middle-aged man screaming at a younger Black young adult, perhaps a teenager. The older man was incredulous about how his whole life had been spent fighting the police with anger and rage and it had got him nowhere. With seemingly blind fury, the man spoke directly to the younger man about how their generation needs to act differently and not repeat the mistakes of his generation as it would only repeat the cycle of violence all over. His voice cracked and tears streamed down his face. The young man stood stoically but nodded and seemed utterly enraptured. The older man clearly was hurt, angry and desperate. In many respects, it was not the typical scene of love at all. However, his righteous anger was trying to bridge the generational divide and impart lessons from decades of Black oppression to the younger cohort. It was an educational tirade but one that attempted to construct a new reality for Black people in America.

Another video that went viral was of a line of protesters facing armed police. The police start to advance. A young Black man breaks ranks from the protest line to kneel in front of the advancing police. Seeing that the police are drawing their batons with seemingly every intention of violently attacking this young protestor, a white girl, perhaps even younger, steps in front of the protester and turns her back on the police. As the armed police continue to march forward they press up against the girl, who presses up against the boy. They become encircled but the police do not physically assault either of them (at least not before the video ends). The young girl recognized her privilege as a white young woman, whom the police would no doubt be less likely to attack (and the media quicker to defend), to protect this young Black man, who would have more than likely suffered police brutality had she not intervened.

Beyond these fleeting but nonetheless loving examples, recent years have seen mass mobilizations of people to

Ethic 7: Love

help construct and maintain refugee camps in defiance of the state; acts of communal solidarity in the face of ultra-violent religious fundamentalism (from both East and West); personal sacrifices to support the unknown neighbour in need; public workers forgoing personal safety to retrieve immigrants and working-class people from burning buildings made precarious by neoliberal urban governance; mutual aid networks that respond to vulnerable people in a pandemic while the state slowly and inefficiently grinds into action: there are countless stories of an active, unconditional love that goes 'unrewarded' or is even punished by authorities or hegemonic forces looking to quell the revolutionary power of *agapē*. But such love energizes the Other via a unification of power; it untethers the agency of the marginalized and oppressed (albeit perhaps temporarily) from simply having to survive, and aids in the formulation of that agency as a revolutionary force.

The political philosopher Michael Hardt has written extensively about love as a political force, but too often renders any act of love that is 'unjoyful' or angry as not a constructive political force. I beg to differ. Because what of those bodies and people that are in danger when loving unconditionally? What of oppressed communities who could be killed for expressing a genuine *agapē* love? What of LGBTQ+ people who could be criminalized and murdered by a homophobic state? What of faithful and spiritual devotees who have to worship in private due to fear of a religiously extreme and genocidal government? And what of the racialized Other who everyday experiences micro-aggressions in the workplace, the street, the waiting room, public transport, and even at home? Can these bodies really be expected to act out an unconditional love? And even more controversially, what about a love for your oppressors? Is it really possible to love someone who wants you, your body and your way of life destroyed?

As difficult as it is, the short answer is yes. An *agapē* love is not all-out joy; it is not 'sentimental and anaemic'; it is not a naïve optimism that blindly fails to recognize structural and social differences, oppression and violence, as Hardt has argued it is. *Agapē* forgoes the comforting ground that has been built by systems of privilege, tearing those systems down, and reallocating the privilege and resources more equitably. This will inevitably led to discord, debate, factional infighting and dissent. Intersectional and solidarity politics that this entails can create schisms and conflict, but, as Krenshaw, Watson and other have shown (as detailed in Ethics 3 and 5 respectively), are vital to progress. This can 'feel' very unlike the joyful picture of love that Hardt likes to paint. If a protest movement becomes fractured, does it fail? This is why *agapē* love is ethical; it embraces this failure and discord because it can breed alternative ways of resistance to capitalist hegemony and of the progression of commoning. As the cultural theorist Max Haiven has argued: 'The horizon of the struggle for the commons is not some transcendental state when all conflict or disagreement or difference has come to a blissful utopian conclusion. Rather, it is to create the sort of society that is open to change and reflexive, that we can structure ourselves through our cooperative efforts.'[15] Decentring ourselves, forcing ourselves to listen to others constructively when we think they are wrong, is tiring, slow and sometimes even dangerous. Which is why nourishment of the self – or self-care – is an important practice of *agapē*. It exists within a suite of other ethics that help sustain as well as action it. It is a resource we expend on but also obtain from others, both human and nonhuman.

Love as *agapē* is revolutionary precisely because it cuts

Ethic 7: Love

across bodies, cultures, races, religions, politics, disciplinary boundaries, discourse, language, affect and thought. It is a love that is reckless in praxis but constructive in deed. It reconstitutes the public sphere from the appropriative clutches of privatization. It takes its shape in radical critique and broader movements such as feminism, queer studies, decolonization, disabled and anti-racist struggle precisely because it shuns the comforting ground that comes with majoritarian hegemonic power. This refusal to pin the theorization of love to any discipline or hierarchical mode of thought renders the benefits of this radical love elusive to the central powerful majority, and provides an ethical resource for those at the exploited margins. Rather than fighting for empowerment, *agapē* fights against disempowerment in the first place. Rather than seeking diversity, it questions why we need to be divergent at all.

In a world characterized by increasing hate, one in which the languages of genocide have crept back into the mainstream through a political-media industrial complex too obsessed with holding on to power, talking about this radical praxis of love – one that actively and unconditionally loves your neighbour (wherever she may be) – is more critical than ever.

This reading of love fuels the realization of a planetary commons. It *is* the ethical nature of the communing process that is infectious and spreads from person to person, from community to community. *Agapē* politically necessitates a common resource to be first and foremost utilized by the community it serves. And it offers a powerful praxis by which that community can self-manage the resources they require. Commoning is the act of resisting capitalist accumulation and replacing it with a democratic management and equitable allocation of shared resources. This is not possible without an ethical commitment to mutualism, transmaterialism, minoritarianism, decodification, slowness, failure and love. But to paraphrase Saint Paul, the best of these seven is love.

Conclusion:
The State of the Commons

Throughout this book, I have sought to navigate a conceptual path towards a planetary commons. Using already-existing examples of commoning and anti-capitalist praxis, an ethics of how that planetary commons can be realized comes into view. The seven ethics I have outlined in this book, if thought about collectively, and also actively and politically, have the potential to radically reshape our society and bring a fairer, more just world sharply into focus. But how would this work in the real world? The capitalist realism of the twenty-first century seems irrevocably immutable; surely it is too late to create a different, more common world order? I am under no illusion that it will be easy, particularly as the real and present catastrophe of climate change is radicalizing our governments into further fascistic behaviour. However, what would a future that managed to turn the tide look like? If you will allow me to dream for a moment, what would a world look like that took the seven ethics I have outlined seriously, and tried to apply them within the very *unethical* frameworks of institutional politics? How would they percolate through, and perhaps clash with, a desire from the powerful to govern? Would an ethically and planetary common world ever be compatible with our current version of electoral politics in the global north? What about the work needed to undo the rampant violation of the global south? With all that in mind, perhaps this is how the next decade might pan out . . .

— *Happy New Year!* —

There are going to be plenty of these 'New Year's Resolution' pieces in the next few days, particularly as we enter a brand new decade. The tumultuous twenties are ending, and perhaps, just maybe, we're entering the thriving thirties? Because for the first time in as long as I can remember, I am not looking into the future with existential dread.

There are plenty of examples of really important events that have happened in 2029 that we can use as springboards to carry us on through into 2030 and beyond: the implementation of universal basic services, digital dividends, rent control in our global cities, the four-day working week and free augmented-reality contact lenses for the over fifty-fives. And we can look forward to the global reparations bill being rubber stamped at the United Nations in the next few months, which will give the countries of the global south, those most affected by climate change, the resources they need to act.

But none of these would have been possible had it not been for the wake-up call of the coronavirus pandemic of 2020. For nearly two years, as the world experienced lockdown after lockdown as the virus spiked and re-spiked, it became obvious to all and (better late than never) to the world leaders of the time that a radical change was needed to the way our society was organized. It took the first four

Conclusion: The State of the Commons

years (2024–8) of President Alexandria Ocasio-Cortez's (or PAOC's as she has become known) term in office to reverse all the damage Trump and then the dithering Biden did to the US constitution. It is clear now how close the US came to all-out civil war in 2025 after PAOC signed an executive order that tried Donald Trump and Mike Pence for treason for their subversion of the US constitution throughout Biden's lame-duck presidency of 2020–4. If it hadn't been for the swift action of the US military to stop the Trump family arming the Confederacy Militia Army with weapons of mass destruction, the 'U' in the US might well have had to be erased.

But now, two years into PAOC's second term, it has become clear that we would not be where we are today had it not been for the radical changes brought about by the coronavirus pandemic. It tore a gaping hole in the existing social, cultural and political world, and while it took us a few years to realize this, it offered us the tools – in the form of political will, emotional consciousness and economic resources – to finally tackle climate change head on, and put an end to the right-wing fascistic governments that it was enabled by.

And as we head into the fourth decade of the twenty-first century, CO_2 levels have plateaued in the previous twenty-four months and are forecast to go down in 2032. Yes, DOWN. This coming year is widely predicted to be the first since the 1970s when the Arctic sea ice will actually recover this winter. Oceanic acidification is slowing, and the report from the Berlin Institute of Protein Farming of the first sighting of an insect from the *Bombus* genus may well turn out to be true. Our climate may well be back on track to pre-Covid levels, and from there, who knows? Of the seismic shifts that have happened in the world in the past ten years or so since the Covid-19 pandemic shocked the neoliberal capitalist world order to its core, which is the one that we can point to that brought us back from the brink of climate catastrophe?

The Green New Deal, spearheaded by PAOC, will be seen by the majority as the catalyst for sure – and to some extent rightly so. It wasn't just a recalibration of existing industrial production systems such as waged labour and globalized commodity chains to be more 'green'; it was a wholesale change in how we extracted resources out of the 'natural' world around us. Utilizing the trillions of dollars leveraged in the groundbreaking 'Robin Hood Tax' that targeted financial transactions first, but soon scooped up 90 per cent of all inherited wealth transfers (masterminded by the French minister of the economy and finance Thomas Piketty), the US was able to offer every citizen a basic income, universal healthcare and the ability to 3D print their home. This was a technology that people were slow to take up at first, but when the ability to use waste packaging and old technological devices as the raw material became clear, it enabled people to reuse their unwanted products almost instantaneously.

Politically, there was strict devolution of planning laws that were democratized as far as possible via participatory budgeting, civic assemblies and online voting apps, which gave local people the final say on how community resources were to be allocated. Reparations to the Black and indigenous communities gave them the opportunities and financial ability to be involved in these decisions. PAOC has been rightly lauded for this, but some, including ex-President Arce of Bolivia, are pushing for all countries to do what Bolivia did back in 2025, namely completely dissolving the federal government and creating a National Citizens' Assembly to run the country. The inaugural Summit of South American Socialist Leaders (or SSASL

Conclusion: The State of the Commons

for short) in Cherán, Mexico, later this year will be vital if Arce is to convince world leaders they are no longer needed, and to give up their power to their people.

The advances in solar and wind power meant the ban on fossil fuels affected none but the most stubborn of petrol-heads. The 2025 riots in Austin, Texas, by disgruntled oil workers were damaging, yes, and rumours of Russian involvement are still being denied by the now disgraced former Emperor Putin. But with NASAX (the newly formed merger of NASA and Space X) providing a pathway to outer space to collect the minerals needed for building materials from asteroids, there hasn't been a barrel of oil produced in the US since 2027. And the electrification of the second phase inter-state highway loop system that will connect the US to Latin America is due to be finished in 2035, and will finally allow a smoother flow of workers and families between North and South America.

In the United Federation of Europe (UFE), President Annalena Baerbock (formerly the Green Party chair of Old Germany) spearheaded the move to make all workers members of their companies' profit schemes. Emulating the co-operative model of the late 2010s, it meant that every worker had a say in how the company was run. This was chaotic at first. When Volkswagen collapsed after the famous emissions scandal of 2024 (but then, they did have previous . . .), it threatened to derail the co-operative model. But the new UFE-wide 'degrowth' law of 2025 (modelled on New Zealand's highly successful model of 2019), which mandated putting worker well-being and ecological sustainability above profit-making as corporate goals, eradicated the incentives for corporate corruption almost overnight. The purge of the greedy CEOs was swiftly over, unlike their jail sentences. The four-day working week allowed more time for rest and as it produced far higher mental health scores than ever before, it actually made workers far more productive. And the 'No-Tech Tuesdays' movement lead by Professor Carole Cadwalladr is likely to become UFE policy before too long.

Legally too, there was a drive to enact the Mother Earth laws that originated in Bolivia in 2012, and proliferated under President Arce in the early twenties. Rivers, mountains, fields and forests across Latin America were given the same legal status as people, and the first conviction for the murder of a forest by a Brazilian logging firm in 2026 sent shock waves through the boardrooms of extraction industry companies – so much so that the now heavily regulated mining on that continent is at its lowest levels in a century. PAOC has recently enacted a similar law for Yellowstone Caldera (although this was an attempt partly to protect its volcanic volatility), and so the hope is that across the global north, other countries will begin to recognize 'natural' landscapes as living entities in their own right.

But it was perhaps PAOC's government's recalibration of food production that did the most to halt the runaway carbon emissions of the agricultural industry. Requiring community gardens to be set up for every new neighbourhood built, and the growing of food from electricity to be taught to every school child, meant that we now have a generation of people who eat on average 75 per cent of their diet from their local neighbourhood garden, according to the latest app-generated social indicator scores.

Indeed, those scores would not be possible without the Kiberan Revolution of 2023. Once Africa's largest slum, Kibera in Nairobi housed thousands of workers who did the bidding of Silicon Valley's most exploitative tech companies. Working for next to nothing to provide all the data for the artificial intelligence (AI) systems of (the

Conclusion: The State of the Commons

now defunct) Amazon that made obscene profits for their former CEO Jeff Bezos, it is little wonder that the workers revolted. But more than simply demanding better pay and a share of the profits their labour created, the tech-savvy female entrepreneurs demanded a digital dividend to all citizens of Kibera, not unlike the Alaskan Oil Dividend that had been in place since 1976 (mothballed in 2026 once oil production ceased). They argued that the resources of the slum that were going towards producing all the data for the AI systems (such as those used in Uber's driverless cars in Silicon Valley and Amazon's delivery drones across Europe) should be first and foremost controlled by the community. The subsequent bitter international court battle saw Bezos forced to hand over control of his AI systems to not only the Kiberan workers, but every employee who contributed. He still made upwards of a billion dollars a day, but that was not enough for him, so he joined Elon Musk and the other hyper-capitalists in cryostasis who wait for the day when the 'socialist dystopia' of the contemporary world comes to an end.

This digital dividend policy spread around the world and now all content producers – be they a Generation Alpha TweetBook user or an old-school Gen Z TikToker – have the opportunity to control and, crucially, create the platforms that disseminate their content. With the international digital commons laws of 2027 put forward by the UFE's highest citizen assembly, all personal tech platforms must be backwardly compatible for a minimum of ten years and adhere to the stringent 'plug out' test (an innovation devised by the Hyperhumanist think tank).

The tumultuous twenties of the twenty-first century were perhaps as bad as the twenties of the twentieth. But with the foresight of key global leaders including PAOC, President Arce, Prime Minister Ardern and President

Baerbock as well as the International Committee for Disability Foresight (which makes all new policies compatible with disabled people first and foremost), for the first time I am bringing in a new year with hope. Out of the ashes of a burning world fuelled by the hatred of Trump, Johnson, Farage, Bolsanaro, Orbán and all the other hyper-nationalistic leaders of the 2020s, we have all helped to build a world in common. Maintaining it won't be easy for sure; there will always be people looking to hoard the common wealth as their own and dismantle the fragile communities that we have painstakingly built to manage them. But with the political will and, importantly, an ethical commitment *to* that political will from within our own communities, we can begin to look forward to the future instead of dreading it. And that is my new year's resolution. Happy New Year!

An ethical summary

The ethics described in this book – mutualism, transmaterialism, minoritarianism, decodification, slowness, failure and love – are the weapons needed to fight for a planetary commons out of the rampant capitalism that is devouring our world. The kind of world described above is utopian for sure (in the main), but I have used examples from the world that exist already and, if only they are amplified ethically, offer a tantalizing way through the current climate crisis.

These ethics fight for, and outline, a planetary commons that is not simply a single natural resource protected from human use, but a living community of people and things that sustain the flourishing of life. They offer the fuel to spread this ethical praxis across the planet to encompass all people and things in all their diversity. Each ethic in and of itself will not suffice. They may well be progressive ideals and allow people to individually escape the destructive power of capitalist processes for a period of time. They may even allow groups of people to live a life free of capitalist desires completely. But without each other, these ethics will never punch through the suffocating capitalist ideology that haunts everyone and everything in this world. They are ethical precisely because they are faithful to the progressive truths that have been given new life by the (potential) event of the coronavirus pandemic.

This is because the global tumultuousness the Covid-19 event imprinted on the human and nonhuman world has shown how vulnerable capitalism is and how its foundations are not as sturdy, and its roots not as deep, as those in power would have everyone believe. Being ethical going forward is to stay faithful to this truth; to continue to act and construct our world in the knowledge there isn't a capitalist realism and that a more just, equitable and ecologically sustainable post-Covid world is possible.

Specifically, each ethic has outlined the *collective* commoning practices that are needed to propel ethical thinking towards a planetary commons. It is why mutualism is the first and foremost ethnical commitment because without it, any subsequent attempt to build a commons that encompasses a planetary system is doomed to fall. Without decentring the self and shunning the rampant self-interest that is the cornerstone of capitalism, an alternative system will never be built. The other ethical dispositions flow from this mutualism. This is why, throughout the ethics, their collective and indeed their structural characteristics have been emphasized. As has been described, capitalism's appropriative protocols will easily co-opt any individualized ethical commoning behaviour and reroute it towards its own ends.

This is because capitalism utilizes the existing structures at its disposal – political systems and parties, the media, educational systems and of course corporations – to do its bidding. We saw that most acutely with the ethics of both transmaterialism and slowness. With the former, ethical consumerism has been introduced by many corporations, from which many of us rush to buy to assuage our guilt, but without changing the ecocidal processes embedded with their production. And with the latter, capitalist structures and corporations can use self-help as a way to manage their constituents, all the while prepping them as more productive, resilient and hence exploitable workers. The lure of capitalism's stardom, excitement and riches for those willing to step *out* of the collective networks of solidarity as individualized champions of progress can take hold at any point of the commoning process. So with all seven ethics, their collective functioning has been emphasized and their structural capabilities put forward. Because without these, a future beyond capitalism will never be built.

Hence, this structural collectivity is a prerequisite for a planetary commons precisely because it helps to narrate the community aspect of the dialectical understanding of the commons put forward by Gibson-Graham and Gudeman (as outlined in the introduction). The interplay between the community and a common resource enlivens the commons as a functional ideology of the socio-economic world. It rejects the institutionalized 'safeguarding' that comes about from viewing the commons through the lens of supra-national institutions such as the United Nations, or as Ostrom's articulation of the commons might afford. The community's historic, present and future engagement of a common resource (be that a forest, river basin, community urban garden, virtual space, digital encyclopaedia) enlivens it and makes the content available to those who need it most. It is the *ethical* application of a common resource. And as Elias and Moraru argued in their articulation of planetarity, such an ethical application creates infectiousness and a virality that is contagious and aids in the diffusion of not only the resource but, critically, the ethical commitments that sustain it. Covid-19 has shown us just how connected and dependent on our sociality we are as a globalized community; using that same essence of virality to spread ethical behaviour instead of disease must be part of a commoning praxis.

However, a note of caution is needed. The ethical dispositions that have been offered in these last seven chapters, in a rather cruel twist of fate, are countered by this book simply existing. For starters, there are seven of them, when according to Ethic 4, quantification is to be resisted. In addition, there is my name on the front cover, cutting off all the other people and things that have contributed to my writing it.[1] And I will no doubt benefit institutionally via the existing structures of higher education if this book is 'successful' in the traditional sense of the word. It is written in English, the world's major language, and so is immediately unreadable by a vast swath of the planet's population; and compared to some academic publishing protocols, it has been a relatively quick process. The publishing industry too is not immune from accusations of environmental damage; the pages (or screen) that you are reading this on will have been extracted from a natural resource that is undoubtedly a part of an exploitative production chain. The very act of articulation-by-publication of these ethics risks them slipping into the world as a co-optable 'thing'; an ideology or process that can be attacked by capitalism's foot soldiers of stardom, spectacle and short-termism. The hope is that the resultant commoning processes that this book catalyses outweighs any co-option that occurs.

There is also the important social categorization of class that shoots through many of these ethical concerns about the collective. Working-class solidarity and the trade union movements that it engendered have been campaigning for many decades for the power of the collective to resist the individualizing power of capitalism and corporate practices. In addition, the living conditions of the poorest of us across the world are imbued with these ethical dispositions already, but out of the need for survival rather than any sense of forcibly growing a planetary commons in reaction to capitalism. Favela dwellers, garbage dump inhabitants, the homeless, council estate kids: they will often have to rely on others, the material world around them, slowness and failing far more than anyone else. The homeless, for example, are often some of the most transmaterial among city dwellers, being forced to use whatever nonhuman material they can lay their hands on for warmth, food or simple survival. Being a vegan may be an important lifestyle choice, but only if you can afford it. And council estate kids in the UK (and across the global north) have often been stigmatized for being anti-social while they embody a slowness by simply 'hanging out' with each other in public places because their youth club has been closed down in the most recent rounds of local government cuts. Of the examples and characteristics of the ethics outlined in this book, many are already employed socially; it is just that because they are employed by the working and/or underclass in both the global north and south, they are stigmatized as anti-social, problematic, backward and generally not what is needed to progress to a system beyond capitalism. This could not be further from the truth. Because of the ethic of minoritarianism, these people's existing experiences of commoning will be vital to learn from and input into larger political narratives. Along

Conclusion: The State of the Commons

with Black and ethic minority, queer and disabled voices, the working class are too often marginalized and ignored in progressive narratives.

There is also a multitude of other avenues of exploration that fly off from the ethical framing of this book. For example, what of the role of art in the commons? Artistic production within capitalism serves a dualistic purpose: there is art that contributes to the swelling of the bottom line (co-opted creative practices) and that which subverts the capitalist agenda; there is very little room for much in between, such is the insatiability of capitalist's appropriative mechanism. But thinking about how art plays a role in championing the ethical mindset outlined in this book will be critical in diffusing it through communities and societies. The conceptualization of art within a planetary commons will be vastly different from the binary nature it is beholden to under capitalism; but that, as they say, is for another day.

Capitalism has ushered in the climate catastrophe and untold injustice on a global scale. We are in desperate need of a radically different way of organizing a world-in-common that is fairer, more just, democratic and equitable; we need a *planetary commons*. If this is to be realized then we – the global multitude of the human race in all its glorious difference – need to act ethically with and through the human and nonhuman world around us. This book has been a journey through some of the existing ways in which people and communities are trying to envision a world beyond capitalism. From these, ethical commitments to the commoning and anti-capitalist truths that have been revealed by the coronavirus-evental rupture in neoliberal-cum-fascistic capitalism have been articulated. The book has outlined seven, but there may be more, there may be less. They are not a framework for a future world that must be adhered to, nor a blueprint for commoning. These ethics are a way to simply keep that rupture going, to continue the ways of questioning, resisting and admonishing the unjust and climate-destroying structures of a capitalist way of life. In so doing they open up and enliven a future horizon of commonality; one that is universal and planetary.

By drawing on the threads of equity, justice, activism and resistance, the fabric of a common socio-economic world can be woven. It will not be easy, as the forces of capitalism that look to maintain the status quo of limiting the abundant benefits of this world to a carefully selected few are overwhelmingly powerful. But commoning – if performed with mutualism, transmateriality, minoritarianism, decodification, slowness, failure and love; in short, *ethically* – is far more so.

Notes

Introduction

1 Spivak, G. (2014) Planetarity. in Cassin, B. (ed.) *Dictionary of Unstranslatables: A Philosophical Lexicon*. University of Princeton Press, Princeton.
2 Holloway, J. (2010) *Crack Capitalism*. Pluto Press, London.
3 Mould, O. (2018) *Against Creativity*. Verso, London.
4 Harvey, D. (2010) *Rebel Cities*. Verso, London.
5 Gibson-Graham, J. K. (2006) *A Postcapitalist Politics*. University of Minnesota Press, Minneapolis. For those of you unfamiliar with Gibson-Graham, this is the singularized pen name for the geographers Julie Graham and Katherine Gibson.
6 See Greer, A. (2012) Commons and enclosure in the colonization of North America. *The American Historical Review*, 117(2): 365–86; and Fortier, C. (2017) *Unsettling the Commons: Social Movements Within, Against, and Beyond Settler Colonialism*. ARP Books, New York.
7 Gudeman, S. (2001) *The Anthropology of Economy: Community, Market, and Culture*. Blackwell, Oxford, p. 27.
8 Federici, S. (2018) *Re-Enchanting the World: Feminism and the Politics of the Commons*. PM Press, New York, p. 110.
9 Baland, J. M., Bardhan, P., Das, S. and Mookherjee, D.

(2010) Forests to the people: Decentralization and forest degradation in the Indian Himalayas. *World Development*, 38(11): 1642–56.
10. Ellickson, R. and Thorland, C. (1995) Ancient land law: Mesopotamia, Egypt, Israel. *Chicago Kent Law Review*, 71: 321–408.
11. Hegenman, S. (2019) The indigenous commons. *Minnesota Review*, 93: 133–40.
12. Winstanley, G. (1652) *The Law of Freedom in a Platform*. Available freely online.
13. Reclus, E. (1871) quoted in Ross, K. (2015) *Communal Luxury: The Political Imaginary of the Paris Commune*. Verso, London, p. 5.
14. Mould, O. (2020) Revolutionary ideals of the Paris Commune live on in Black Lives Matter autonomous zone in Seattle. *The Conversation* online.
15. Ostrom, E. (2015) *Governing the Commons: The Evolution of Institutions for Collective Action*. Cambridge University Press, Cambridge.
16. Bollier, D. (2014) *Think Like a Commoner: A Short Introduction to the Life of the Commons*. New Society, New York, p. 6.
17. Hardt, M. and Negri, A. (2009) *Commonwealth*. Harvard University Press, Cambridge, MA, p. 139.
18. I have used these examples of the commons to exemplify my ideas, but there are many more that could have been used.
19. For an overview see the collection edited by Micocci, A. and Di Mario, F. (eds) (2017) *The Fascist Nature of Neoliberalism*. Routledge, London.
20. Wilkinson, R. and Pickett, K. (2018) *The Inner Level: How More Equal Societies Reduce Stress, Restore Sanity and Improve Everyone's Well-Being*. Penguin, London.
21. See the work of anarchist geographer Simon Springer, notably Springer, S. (2016) *The Anarchist Roots of Geography: Toward Spatial Emancipation*. University of Minnesota Press, Minneapolis.

22 Lovelock, J. (2007) *The Revenge of Gaia: Why the Earth Is Fighting Back – and How We Can Still Save Humanity*. Allen Lane, Santa Barbara.
23 Latour, B. (2017) *Facing Gaia: Eight Lectures on the New Climatic Regime*. Polity, Cambridge.
24 Ibid.
25 Elias, A. and Moraru, C. (2015) *The Planetary Turn: Relationality and Geoaesthetics in the Twenty-First Century*. Northwestern University Press, Boston, pp. xi–xii, original emphasis.
26 Deleuze, G. and Guattari, F. (1988) *A Thousand Plateaus: Capitalism and Schizophrenia*. University of Minnesota Press, Minneapolis.
27 Loraine, T. (2011) *Deleuze and Guattari's Immanent Ethics: Theory, Subjectivity, and Duration*. SUNY Press, Albany, p. 1.
28 Gibson-Graham, *Postcapitalist Politics*, p. xxxi.
29 Deleuze, G. (2004 [1969]) *The Logic of Sense*. Continuum, London, p. 163.
30 Ibid., p. 191.
31 Badiou, A. (2012) *The Rebirth of History: Times of Riots and Uprising*. Verso, London.
32 PA Media (2020) Wealth tax on rich should aid UK's Covid-19 recovery, says Labour. *Guardian* online.
33 Spade, D. (2020) *Mutual Aid: Building Solidarity During This Crisis (and the Next)*. Verso, London.
34 Roy, A. (2020) The pandemic is a portal. *Financial Times* online.
35 Markovčič, A. (2020) Capitalism caused the COVID-19 crisis. *Jacobin* online.
36 Solnit, R. (2020) 'The impossible has already happened': What coronavirus can teach us about hope. *Guardian* online.
37 For example, the UK right-wing commentator Toby Young's article in the *Critic*, March 2020, which I advise you not to go and read lest he be encouraged.

38 Cook, J. (2020) Netanyahu uses coronavirus to lure rival Gantz into 'emergency' government. *CounterPunch* online.
39 Lent, J. (2020). Coronavirus spells the end of the neoliberal era. What's next? *OpenDemocracy* online.
40 Gardner, K. and Clancy, D. (2017) From recognition to decolonization: An interview with Glen Coulthard. *Upping the Anti*, 19, online.

Ethic 1

1 Whether he read it himself, or had it read to him, though, is unclear.
2 Freeland, J. (2017) The new age of Ayn Rand: How she won over Trump and Silicon Valley. *Guardian* online.
3 It is worth pointing you here towards the work of C. B. Macpherson, who has detailed far more eloquently the differences in the philosophies of individualism of each of these Enlightenment thinkers, notably in his 1962 book *Possessive Individualism* (Oxford University Press, Oxford).
4 Smith, A. (1776 [2000]) *The Wealth of Nations*. Modern Library, New York, book IV.ii, p. 9.
5 Hayek, F. (2001 [1944]) *The Road to Serfdom*. University of Chicago Press, Chicago, p. 8.
6 Hayek, F. (1968) Der Wettbewerb als Entdeckungsverfahren. Lecture at the University of Kiel.
7 Brown, W. (2015) *Undoing the Demos: Neoliberalism's Stealth Revolution*. Zone Books, New York, p. 31.
8 Hall, S. (1980) Thatcherism: A new stage? *Marxism Today*, 24(2): 26–8.
9 Klein, N. (2007) Disaster capitalism: The new economy of catastrophe. *Harpers*, October: 47–58.
10 Quoted in Peron, J. (2015) There is nothing libertarian about conservatives. *Huffpost*, 17 March.
11 See Monbiot, G. (2018) *Out of the Wreckage: A New Politics for an Age of Crisis*. Verso, London; and the many musing of Joseph Stiglitz.

12 Kropotkin, P. (2012 [1902]). *Mutual Aid: A Factor of Evolution*. Courier, London.
13 Springer, S. (2016) *The Discourse of Neoliberalism: An Anatomy of a Powerful Idea*. Rowman & Littlefield, New York.
14 Iacoboni, M. (2009) Imitation, empathy, and mirror neurons. *Annual Review of Psychology*, 60: 653–70.
15 Rifkin, J. (2010). *The Empathic Civilization: The Race to Global Consciousness in a World in Crisis*. Polity, Cambridge.
16 Anderson, B. (2006). *Imagined Communities: Reflections on the Origin and Spread of Nationalism*. Verso, London.
17 Prøitz, L. (2018) Visual social media and affectivity: The impact of the image of Alan Kurdi and young people's response to the refugee crisis in Oslo and Sheffield. *Information, Communication & Society*, 21(4): 548–63.
18 Mould, O. (2018). The not-so-concrete Jungle: Material precarity in the Calais refugee camp. *cultural geographies*, 25(3): 393–409.
19 Savage, D. C. (1977). Microbial ecology of the gastrointestinal tract. *Annual Review of Microbiology*, 31(1): 107–33.
20 Lederberg, J. and McCray, A. T. (2001) 'Ome sweet 'omics: A genealogical treasury of words. *The Scientist*, 15(7): 8.
21 Alcock, J., Maley, C. and Aktipis, C. (2014) Is eating behavior manipulated by the gastrointestinal microbiota? Evolutionary pressures and potential mechanisms. *Bioessays*, 36(10): 940–9.
22 Conniff, R. (2013) Microbes: The trillions of creatures governing your health. *Smithsonian Magazine* online.
23 Gangopadhyay, A., Srivastava, A., Srivastava, P., Gupta, D., Sharma, S. and Kumar, V. (2010). Twin fetus in fetu in a child: A case report and review of the literature. *Journal of Medical Case Reports*, 4(1), art. 96.
24 Kramer, P. and Bressan, P. (2015) Humans as superorganisms: How microbes, viruses, imprinted genes, and other selfish entities shape our behavior. *Perspectives*

on Psychological Science, 10(4): 464–81, p. 475, my emphasis.
25 In *A Thousand Plateaus*, Deleuze and Guattari write of a full, empty and cancerous BwO, and as such, a BwO that can overspill into uncontrollable desire; they use the example of drug addicts who pass the healthy BwO stage and career off into death.

Ethic 2

1 Simon, J. (2013) *Neomaterialism*. Sternberg Press, Berlin.
2 Steyerl, H. (2017) *Duty Free Art*. Verso, London.
3 Simon, J. (2010) Neo-materialism, part I: The commodity and the exhibition. *e-flux*, 20, online.
4 I have expanded on this elsewhere: see Mould, O. (2019) The spark, the spread and ethics: Towards an object-orientated view of subversive creativity. *Environment and Planning D: Society and Space*, 37(3): 468–83.
5 Yusoff, K. (2018) Politics of the Anthropocene: Formation of the commons as a geologic process. *Antipode*, 50(1): 255–76, p. 270.
6 Yusoff, K. (2018) *A Billion Black Anthropocenes or None*. University of Minnesota Press, Minneapolis.
7 This is why, Yusoff argues, the construction of race is inherently bound up with the history of the extraction economy; 'racialization belongs to a material categorization of the division of matter (corporeal and mineralogical) into active and inert' (ibid., p. 2).
8 Veltmeyer, H. and Petras, J. (2014)*The New Extractivism: A Post-Neoliberal Development Model or Imperialism of the Twenty-First Century?* Zed Books, London.
9 To use the language of Jane Bennett, discussed later.
10 The prefix 'trans' here is a deliberate usage, because of its transcendental qualities, but also because of the struggles of the transgender community. To be 'trans' in this instance is to be between the two binary genders, either

transitioning from one to the other, or in a settled gendered identity between the two. The transgender identity hence problematizes the traditional gender duality and opens up new ways of engaging with human identity, and frees people from the oppressive enclosure such binary thinking can sometimes have. I use the term 'transmaterialism', then, in part in solidarity with the oppressed trans community, but also as I am aiming for similar emancipation-from-duality potentials.

11 Food and Agriculture Organization of the United Nations (n.d.) *Key Facts and Figures*, online.
12 Francione, G. (2008) *Animals as Persons: Essays on the Abolition of Animal Exploitation*. Columbia University Press, New York, p. 16.
13 The intersectionality of these movements is something that will be discussed in the next ethic.
14 Francione, *Animals as Persons*, p. 17.
15 Blytham, J. (2016) Can hipsters stomach the unpalatable truth about avocado toast? *Guardian* online.
16 Groups such as Brandalism engage in subvertising and their work can be seen across cities all over Europe and beyond.
17 This encompasses the concept of 'rentier capitalism', as discussed by Standing, G. (2016) *The Corruption of Capitalism: Why Rentiers Thrive and Work Does Not Pay*. Biteback, London.
18 Nixon, R. (2011) *Slow Violence and the Environmentalism of the Poor*. Harvard University Press, Cambridge, MA.
19 Frost & Sullivan (2018) *The Impact of Digital Transformation on the Waste Recycling Industry*. Online.
20 Bennett, J. (2010) *Vibrant Matter: A Political Ecology of Things*. Duke University Press, Durham, NC.
21 This was something that the mutual aid network I was part of undertook once it became abundantly clear that the UK government were failing to procure enough PPE for frontline healthcare staff.
22 Benn, T. (1976) The Levellers and the English democratic

tradition. Speech at the Bertrand Russell Peace Foundation for the Oxford Industrial Branch of the WEA.
23 Yusoff, *Billion Black Anthropocenes*.
24 DeLanda, M. (2016) *Assemblage Theory*. Edinburgh University Press, Edinburgh.
25 Harman, G. (2018) *Object-Oriented Ontology: A New Theory of Everything*. Penguin, London.
26 See Battisoni, A. (2019) Material world. *Dissent Magazine* online, for a very succinct counter-argument to Latour's brand of materiality, which came to characterize his earlier philosophical work with actor-network theory and its insistence upon the agency of nonhumans and the construction of scientific fact.
27 Tola, M. (2018) Between Pachamama and Mother Earth: Gender, political ontology and the rights of nature in contemporary Bolivia. *Feminist Review*, 118(1): 25–40.
28 Aronoff, K. (2020) The socialist win in Bolivia and the new era of lithium extraction. *New Republic* online.

Ethic 3

1 Fisher, M. (2008) *Capitalist Realism*. Zero Books, London, p. 2.
2 Deleuze, G. and Guattari, F. (1987) *A Thousand Plateaus: Capitalism and Schizophrenia*. University of Minnesota Press, Minneapolis, p. 291.
3 Akala (2018) *Natives: Race and Class in the Ruins of Empire*. Two Roads, London.
4 Deleuze and Guattari, *Thousand Plateaus*, p. 470.
5 Mould, O. (2018) *Against Creativity*. Verso, London.
6 Schulman, S. (2011) Israel and 'pinkwashing'. *New York Times* online.
7 Crenshaw, K. (1989) Demarginalizing the intersection of race and sex: A Black feminist critique of antidiscrimination doctrine, feminist theory, and antiracist politics. *University of Chicago Legal Forum*, (1): 139–67.

8 Butler, J. (1993) *Bodies That Matter: On the Discursive Limits of 'Sex'*. Psychology Press, New York, p. 5.
9 *Black Lives Matter* (n. d.), online.
10 Credit Suisse (2016) *The CS Gender 3000: The Reward for Change*. Credit Suisse Research Institute, Zurich, p. 4.
11 This is why there has been pushback recently on the use of the acronym BAME, as it lumps together very different intersecting forms of race and ethnicity unproblematically.
12 Deleuze and Guattari, *Thousand Plateaus*, p. 292.
13 Althusser, L. (2006) Ideology and ideological state apparatuses: Notes towards an investigation. In Sharma, A. and Gupta, A. (eds) *The Anthropology of the State: A Reader*. Blackwell, Oxford.
14 Butler, J. (1997) *The Psychic Life of Power: Theories in Subjection*. Stanford University Press, Stanford, p. 106.
15 Ibid., p. 130.
16 Katz, C. (1996). Towards minor theory. *Environment and Planning D: Society and Space*, 14(4): 487–99, p. 494.
17 Coates, T.-N. (2014) The case for reparations. *The Atlantic*, June. This 'version' of reparations was not without its critics, notably Cornell West, who derided it as the 'neoliberal face of black freedom struggle': see West, C. (2017) Ta-Nehisi Coates is the neoliberal face of the black freedom struggle. *Guardian* online.

Ethic 4

1 Qin, A. (2020) China raises coronavirus death toll by 50% in Wuhan. *New York Times* online.
2 Carding, M. (2020) Government counts mailouts to hit 100,000 testing target. *HSJ* online.
3 Slobodian, Q. (2020) *Globalists: The End of Empire and the Birth of Neoliberalism*. Harvard University Press, Boston, pp. 57–8.
4 This Marxist term forms the basis of Harvey's 'accumulation by dispossession' theory, which has been used

vehemently in many anti-capitalist critiques, particularly in critiquing gentrification. Harvey, D. (2005) *A Brief History of Neoliberalism*. Oxford University Press, Oxford.
5   I have detailed city rankings and their incessant neoliberalization in my previous work, notably Mould, O. (2015) *Urban Subversion and the Creative City*. Routledge, London.
6   Hardy, G. H. (1949) *Divergent Series*. Oxford University Press, London.
7   Ellenberg, J. (2014) Does 0.999... = 1? And are divergent series the invention of the devil? *Slate* online, my emphasis.
8   Doel, M. (2001) 1a: Qualified quantitative geography. *Environment and Planning D: Society and Space*, 19: 555–72, p. 555.
9   Ibid., p. 556.
10 Merry, S. E. (2016) *The Seductions of Quantification: Measuring Human Rights, Gender Violence, and Sex Trafficking*. University of Chicago Press, Chicago.
11 Ibid., p. 21.
12 Ibid., p. 5.
13 Ajana, B. (2017) Digital health and the biopolitics of the quantified self. *Digital Health*, 3: 1–18.
14 Davies, W. (2015) *The Happiness Industry: How the Government and Big Business Sold Us Well-Being*. Verso, London.
15 Deleuze, D. (1990) Postscript on the societies of control. Online, p. 4, my emphasis.
16 Preciado, P. (2020) Learning from the virus. *Artforum* online, unpaginated.
17 Ibid., my emphasis.
18 Lowrie, W. (1997) *Fundamentals of Geophysics*. Cambridge University Press, Cambridge.
19 König, S. U., Schumann, F., Keyser, J., Goeke, C., Krause, C., Wache, S., Lytochkin, A. et al. (2016) Learning new sensorimotor contingencies: Effects of long-term use of sensory augmentation on the brain and conscious perception. *PLoS One* online.

20 A phrase coined by the computer scientist Carl H. Smith.
21 Wilcockson, T. D., Osborne, A. M. and Ellis, D. A. (2019) Digital detox: The effect of smartphone abstinence on mood, anxiety, and craving. *Addictive Behaviors*, 99, art. 106013.
22 Deleuze, Postscript, p. 6.
23 Quoted in Kiger, P. (2014) Hudson Yards rises above the rails. *Urbanland* online.
24 Mattern, S. (2016) Instrumental city: The view from Hudson Yards, circa 2019. *Places* online.
25 Hawkins, A. (2019) Alphabet's Sidewalk Labs unveils its high-tech 'city-within-a-city' plan for Toronto. *The Verge* online.
26 Gabrys, J. (2014) Programming environments: Environmentality and citizen sensing in the smart city. *Environment and Planning D: Society and Space*, 32(1): 30–48.
27 For the start of this complex and increasingly vast debate see Kitchin, R. (2014) The real-time city? Big data and smart urbanism. *GeoJournal*, 79(1): 1–14.
28 Ash, J. (2015) *The Interface Envelope: Gaming, Technology, Power*. Bloomsbury, London.
29 This can be viewed easily online and I would definitely recommend you take time out from reading this and watch it now; it is less than six minutes long.
30 De Lange, M. (2019) The right to the datafied city: Interfacing the urban data commons. In Cardullo, P., Di Feliciantonio, C. and Kitchin, R. (eds) *The Right to the Smart City*. Emerald, New York.
31 Latour, B. (2020) Bruno Latour: 'This is a global catastrophe that has come from within'. *Guardian* online.
32 Meadway, J. (2020) Creating the digital commons after COVID-19. *OpenDemocracy* online.
33 Papadimitropoulos, V. (2020) *The Commons: Economic Alternatives in the Digital Age*. University of Westminster Press, London.
34 Preciado, Learning from the virus.

Ethic 5

1. Nixon, R. (2011) *Slow Violence and the Environmentalism of the Poor*. Harvard University Press, Cambridge, MA.
2. Boltanski, L. and Chiapello, E. (2005) *The New Spirit of Capitalism*. Verso, London, p. 488.
3. Debord, G. (1970) *Society of the Spectacle*. Black and Red, London.
4. If you've been to a workplace 'mental health booster day' that does nothing but offer meditation classes in some backhand way of trying to boost your productivity rather than actually cure your anxieties, you'll know exactly what I mean.
5. Formerly known as the National Programme for Happiness and Positivity.
6. UAE Federal Government (2020) *Happiness and National Agenda* online, unpaginated.
7. Alter, A. (2017) *Irresistible: The Rise of Addictive Technology and the Business of Keeping Us Hooked*. Penguin, New York.
8. You read that right: seventeen.
9. Çağlayan, E. (2018) *Poetics of Slow Cinema: Nostalgia, Absurdism, Boredom*. Springer, London.
10. Deleuze, G. (1986) *Cinema 1: The Movement Image*. Continuum, London, p. 3.
11. Debord, G. and Wolman, G. (1956) A user's guide to détournement. *Les Lèvres Nues*, 8, unpaginated. Détournement was a technique of critique development by Debord and other Situationists, which involved cutting up maps of Paris and rearranging them in other ways. It was a form of spatial dislocation and disorientation that aided in the subversion of the capitalist Spectacle (or at least, that's what they thought).
12. He of '4'33"' fame – a piece of music in which the whole orchestra set up, sit down and tune up, and then 'play' 4 minutes and 33 seconds of silence.

13 One wonderfully crafted version of this slow scholarship is Keighren, I. and Norcup J. (eds) (2020) *The Landscape of the Detectorists*. Uniform Books, London.
14 O'Connor, S. (2020) Leicester's dark factories show up a diseased system. *Financial Times* online.
15 Fletcher, K. (2010) Slow fashion: An invitation for systems change. *Fashion Practice*, 2(2): 259–65, p. 262.
16 Tanner, K. (2019) *Christianity and the New Spirit of Capitalism*. Yale University Press, London, p. 110.
17 Ibid., p. 114.
18 Ibid., p. 128.

Ethic 6

1 Beckett, S. (1983) *Worstward Ho*. Grove, London, p. 3.
2 FailCon (2020) About. FailCon online.
3 Sandage, S. (2009) *Born Losers*. Harvard University Press, Cambridge, MA.
4 Piketty. T. (2014) *Capital in the Twenty-First Century*. Harvard University Press, Cambridge, MA.
5 Lewis, S. (2014) 'Failure [is] the gap between where we are and where we want to go': A conversation with Sarah Lewis. *Communications Network* online, unpaginated.
6 She apparently said this in a speech to the United Nations in 1985, although she is on record saying that she doesn't like to take the credit for what is collective knowledge.
7 Yes, this is a burn of Keir Starmer's constant performative tweeting of solidarity.
8 Halberstam, J. (2011) *The Queer Art of Failure*. Duke University Press, London, p. 88.
9 Muñoz, J. (2008) *Cruising Utopia: The Then and There of Queer Futurity*. New York University Press, New York.
10 Scott, A. (2008) *Weapons of the Weak: Everyday Forms of Peasant Resistance*. Yale University Press, London.
11 Halberstam, *Queer Art of Failure*, p. 5.
12 Ibid., p. 97.

13 Ross, K. (2015) *Communal Luxury: The Political Imaginary of the Paris Commune*. Verso, London, p. 50.
14 Taken from an English translation of the manifesto available on *Red Wedge*, April 2016.
15 I have focused on this argument in my previous book: see Mould, O. (2018) *Against Creativity*. Verso, London.
16 Haiven, M. (2014) *Crises of Imagination, Crisis of Power: Capitalism, Creativity and the Commons*. Fernbook, Winnipeg.
17 Mould, O. (2020) Revolutionary ideals of the Paris Commune live on in Black Lives Matter autonomous zone in Seattle. *The Conversation* online.
18 Rannila, P. and Repo, V. (2018) Property and carceral spaces in Christiania, Copenhagen. *Urban Studies*, 55(13): 2996–3011.
19 Coppola, A. and Vanolo, A. (2015) Normalising autonomous spaces: Ongoing transformations in Christiania, Copenhagen. *Urban Studies*, 52(6): 1152–68.
20 Harvey, D. (2011) The future of the commons. *Radical History Review*, (109): 101–7, p. 102.
21 See a detailed analysis of the 'Long Live South Bank' campaign in Mould, O. (2015) *Urban Subversion and the Creative City*. Routledge, London.
22 As many Latin American countries that have undergone coups backed by the Central Intelligence Agency (CIA) will testify.
23 Ellsmoor, J. (2019) New Zealand ditches GDP for happiness and wellbeing. *Forbes* online.
24 Halberstam, *Queer Art of Failure*, pp. 120–1.

Ethic 7

1 Corinthians, chapter 13, in *The Message* translation of the Bible.
2 Horvat, S. (2016) *The Radicality of Love*. John Wiley & Sons, London.

3 Said, E. (1983) *The World, the Text, and the Critic*. Harvard University Press, Cambridge, MA.
4 Derrida, J. (2005) *The Politics of Friendship*. Verso, London.
5 Cooper, M. (2017) *Family Values: Between Neoliberalism and the New Social Conservatism*. Zone Books, New York, p. 248.
6 Lewis, S. (2019) *Full Surrogacy Now*. Verso, London, p. 119.
7 Horvat, *Radicality of Love*.
8 Ellul, J. (1986 [2011]) *The Subversion of Christianity*. Wipf and Stock, Eugene, p. 71.
9 Greenman, J. and Schuchardt, R. (2012) *Understanding Jacques Ellul*. Wipf and Stock, Eugene.
10 King, M. L., Jr (16 August, 1967) Address to the Southern Christian Leadership Conference.
11 Nietzsche, F. (2006 [1885]) *Beyond Good and Evil*. Cosimo, New York, p. 13.
12 Tillich, P. (1954) *Love, Power and Justice: Ontological Analyses and Ethical Applications*. Oxford University Press, London, p. 36.
13 Ibid., p. 33.
14 Arendt, H. (1958) *The Human Condition*. University of Chicago Press, Chicago.
15 Haiven, M. (2014) *Crises of Imagination, Crisis of Power: Capitalism, Creativity and the Commons*. Fernbook, Winnipeg, p. 166.

Conclusion

1 I have tried to counter this somewhat by extending the networks of contribution in the acknowledgements, so do read them.

Index

academia, success and failure in 140, 141–2
actor-network theory 73–4
advertising 16, 67
Africa
   Kiberan Revolution 182–3
agapē love 159, 166–76
agriculture 66, 70–1
   farm workers and veganism 63–4
AI (artificial intelligence) 28, 182–3
AirCasting 116–17
Alaskan Oil Dividend 183
All Lives Matter 88
allotments 71
alt-right 88
Althusser, Louis 92
Amazon 118, 183
   'HQ2' 113–14
anarchism 71
   as an ideology 72
   christianarchy 167–8
   see also eco-squats
ancient Greece
   the commons in 11–12
   four words for 'love' in 159–76
   self-interest in 35–6
animals
   animal rights activism 64–5
   and mutualism 45
   petting zoos 137–8
   and veganism 63, 64–5
animism 11
anti-capitalists 71, 117, 136
   anti-capitalist societies 2–3
   commoning 5–6
   and failure 150, 152, 154
Apple 66, 67
Arab Spring 25
Arctic sea ice 179
Arendt, Hannah 171
Aristotle 162
art
   in the commons 189
   financialization of 57–8
   and the Paris Commune 151
   public art and urban codification 113
   queer art 149
artificial intelligence (AI) 28, 182–3
'As Slow As Possible' (Cage) 129–30
Ash, James 115
assemblage theory 74
asylum seekers 83
Athenian democracy 35
Atlas Shrugged (Rand) 33

Index

austerity 43, 83
  and the refugee crisis 49–50
Australia
  The Ghan (television programme) 128–9, 130

Baerbock, Annalena 181, 184
Barcelona
  Can Masdeu 71
Beckett, Samuel 141, 143, 157
Bennett, Jane 68, 70
Berlin
  urban decodification 114
Berlin Wall 44
Bezos, Jeff 37, 183
Bible
  and *agapē* love 158–9, 166–7, 168
Biden, Joe 179
biodiversity 1, 10, 15
biometric data
  and urban decodification 113
biopolitics
  and capitalism 126–7
  and codification 106–9
Black community
  and Hurricane Katrina 138
Black Deaf Lives Matter 93
Black Lives Matter 79–80, 82–3
  and *agapē* love 171–2
  and animal rights 64
  co-option of within capitalism 84
  and intersectionality 86–7, 88, 93
  online platforms 117
  and the Paris Commune 152
  solidarity with 145–6
Black Trans Lives Matter 93
Blair, Tony 43, 44
blindness, people with
  sensory augmentation for 110–11
bodily decodification 109–11
body without organs (BwO) 53–4, 55

Bolivia 180–1
  Mother Earth laws 75, 182
Bolsanaro, Jair 184
Boltanski, Luc 120–3
brain
  mirror neurons 47–9, 50
branding 57
Brazil 98
Bretton Woods institutions 8, 15, 18
British Empire 37
  see also United Kingdom
Brown, Michael 86
Brown, Wendy 40
Burberry 131–2
business ethics 22
Butler, Judith 85–6, 92–3, 94
BwO (body without organs) 53–4, 55

Cadwalladr, Carole 182
Cage, John 129–30
Capital (Marx) 121
capitalism
  and *agapē* love 167, 171
  and art 189
  and codification 99, 100, 106, 108
  commodification and materiality 57–60, 67
  and the coronavirus pandemic 139, 156–7
  and disability 90–1
  and ethical consumerism 62
  and ethics 184, 185–6
  and failure 142, 149, 151, 154, 155–6
  and the family 164–6
  and happiness 124–5
  and minoritarianism 84
  naturalization of 44
  reversal of benefits from 1–2
  and self-interest 37–8
  and sexualized love 160–1
  and slow media 130, 131
  and the slowness ethic 120–3, 126–8, 132, 133–4, 139

Index

capitalism (*cont.*)
  and solidarity 145–6
  and success 141, 148
  throwaway culture 67–8
  twenty-first century 177
  and the wellness industry 137, 138
  see also anti-capitalists; financialized capitalism; neoliberalism
Capitalism and Schizophrenia (Deleuze and Guattari) 121
Capitalist Realism (Fisher) 77–8
Chiapello, Eve 120–3
Chicago School 39
Children of Men (film) 78–9
chimerism 52
China
  coronavirus pandemic 29, 97
  the Cultural Revolution 25
  Han Dynasty 109–10
Christiana commune 71, 152–3, 154, 155
Christianity
  and *agapē* love 166–9
cities
  codification of 'quality of life' in 101
  smart cities 112–16
  urban decodification 116–19
citizenship
  smart citizenship 115–16
class
  and ethics 188–9
classicism 79
climate catastrophe 1, 2, 28, 177, 179–80, 189
  and failure 142, 157
  and the planetary commons 17
  and the slowness ethic 138–9
  and transmaterialism 74
climate change
  and the global south 95–6
  protests 93
climate justice movements 5
Clinton, Bill 42–3, 44
Coates, Ta-Nehisi 95–6

codification 88–109
  and the coronavirus pandemic 97–9
  of happiness 124–5, 127
  and mathematical philosophy 102–5
  meaning of 99–100
  the quantified self 105–9
  smart cities 112–16
  see also decodification
collective sociality 121
colonialism
  and the commons 6, 9
  and minoritarianism 82, 95–6
commodification 57–8
  of love 160–1
  and the 'spirit' of capitalism 120
commoning 5, 6–7, 190
  and ethics 22
  and failure 146, 154–5, 157
  and love 169–70, 174, 175–6
  and materiality 60
  and mutualism 46–7, 54–6
  see also planetary commons
the commons 3, 4–5, 7–10, 30
  history of 11–17
Communal Luxury (Ross) 150–1
communes 71–6
communism, failure of 44, 46
communities
  and common resources 186–7
  decodification 111–19
the compass 109–10
competition
  and neoliberalism 39–40, 41
Confucianism 11
conscientious objectors 168
consumerism 77
  consumer debt 57
  see also ethical consumerism
consumption
  fast fashion 132–3
  and happiness 125
  and the slowness ethic 120, 139
  see also ethical consumption

Index

Cooper, Melinda 165
coronavirus pandemic
  and allotments 71
  as an event 26–31
  and capitalism 139, 156–7
  deaths 97, 98
  and decodification 119
  and ethical behaviour 187
  and failure 155, 156–7
  fast fashion and sweatshops 132
  impact of 2
  lockdown 139, 178
  mutual aid networks 5
  numbers problem 97–9
  pharmacopornographic response to 108–9
  and the planetary commons 17
  and planetary decodification 118
  post-Covid world 178, 179–80, 185
  and the right to repair 68–9
  and the slowness ethic 139
  testing 97–8
  in the United States 98, 155
corporations
  and the coronavirus pandemic 26
  corporate ethics 22
  future reforms 181–2
  and minoritarianism 84, 87
Coulthard, Glen 30
couples and friendship love 166
Crenshaw, Kimberlé 85, 174
Cromwell, Oliver 71
Cuarón, Alfonso 78
cultural differences and diversity
  and capitalism 121–2
cultural institutions
  and ethical minoritarianism 95
cultural internationalism 16, 20
cultural products
  commonly consumed 15–16
culture
  nature/culture divide 18
CVs of failure 140–1, 142

datafication 113
Davies, Peter 163
Davies, William
  The Happiness Industry 124–5
Dayton, Jonathan 146
deaths
  coronavirus 97, 98
  disabled people 89–90
Debord, Guy 129
decodification 32, 97–119, 176, 184, 190
  and *agapē* love 170
  bodily 109–11
  planetary 109, 111, 112, 116–19
  and slow media 130
  urban decodification 112–16
  see also codification
decolonization 175
  and minoritarianism 95, 96
deforestation 9, 10
dehumanization 80
deindustrialization 43
DeLanda, Manuel 73
Deleuze, Gilles
  concept of the 'body without organs' (BwO) 53–4
  Deleuze and Guattari's desire-production 121, 123, 126
  on ethics 22–3, 24
  on events 24
  and minoritarianism 81–2, 83–4, 89, 91, 94–5
  'Postscript on the societies of control' 106–8, 112
  on slow cinema 128–9
  see also Guattari, Félix
democracy
  and friendship 162
  and self-interest 35–6
Denmark
  Christiana commune 71, 152–3, 154, 155
depression 125–6
deregulation 43
Derrida, Jacques 162
desire 121–3, 126, 127
desire-production 121, 123, 126

Index

the Diggers 13, 14, 20, 70–1, 155
digital detoxes 111, 136
digital technology
  and bodily decodification 109, 110–11
  and capitalism 16
  and codification 105–7, 105–8, 112–16
  dating apps 160–1
  the digital commons 18
  and happiness 124
  the 'interface envelope' 115–16
  international digital commons laws 183
  and planetary decodification 116–19
disability 89–92, 184
  'fitness to work' assessments 89–90
disaster capitalism 43
Doel, Marcus 103, 104
domestic abuse, statistics on 104
Donne, John 47
dopamine 124, 125, 126
drug use
  and eco-squats 152
  legalization of marijuana 154
Druids 11

East India Company 37
eco-squats 5, 71–6, 127, 152–5
  Christiana commune 71, 152–3, 154, 155
  Growth Heathrow 71, 72–3, 74, 75, 153
Edison, Thomas 143
educational institutions
  minoritarianism in 95
  university league tables 100–1
Egypt, ancient 11
Elias, Amy 18, 19, 21, 70, 131, 187
Ellul, Jacques 167–8
empathy
  and minoritarianism 92
  and mutualism 47–9, 50, 51, 52–3, 54

England
  seventeenth-century 'commonwealth' ideas 13–14, 70–1
enjoyment 121–3, 125, 127
Enlightenment
  and indigenous peoples 12–13
  and self-interest 35–6
entrepreneurialism 44
environmental costs
  of fast fashion 132, 133
equality
  and minoritarianism 79
equity 190
eros love 159–61, 167
ethical consumerism 69–70, 186
  and veganism 62–3, 64
ethical consumption
  and the right to repair 69
  and veganism 63–5
ethics 21–4, 29–31, 189–90
  codes of 21–2, 100
  of planetarity 21
  of the planetary commons 30, 31–2
  related to events 24
Eurocentrism 83
European Union (EU) 29, 66
#EverydaySexism 83
events 24–6
  the Covid event 26–31
  ethics related to 24
evolutionary theory
  mirror neurons 47–9
exchange value and codification 99–100
excitement
  and the growth of capitalism 121–3
exploitation, abolition of 63, 69
Extinction Rebellion 93

Facebook 166
failure 32, 139, 140–57, 176, 184, 190
  and *agapē* love 170
  as a collective 142

Index

CVs of 140–1, 142
failing 'better' 140, 141
past failures and present actions 134
and revolutionary thought 149, 150–5
and solidarity 144–6
states and ethical failure 155–7
and success 141–2, 146–9, 154
families
family love (*storge*) 159, 164–6
queering success 147–8, 149
Farage, Nigel 184
Faris, Valerie 146
fascism 38, 43, 167
fashion
burning stock 131–2
fast fashion 132, 133
slow fashion 123, 132–4
fast food industry
and veganism 61, 62
Federici, Silvia 9, 15
feminism 165, 175
intersectional 85–6
feng shui 110
the 52 Climate Actions 69
filiation 161, 162
films
Children of Men 78–9
Hyper-Reality 115–16
Little Miss Sunshine 146–9
slow cinema 128–9
financial crash (2008) 43, 44, 57
financialized capitalism
high-frequency trading (HFT) techniques 134–5
Fisher, Mark 77–8
FixMy Street 117
Fletcher, Kate 132–3
flow activities 124, 126
Floyd, George 86
food
growing of 182
'microbiota' and food cravings 51
Ford, Henry 65–6
fossil fuels 59

Foucault, Michel
on the disciplinary society 106–8
The Fountainhead (Rand) 33
Francione, Gary 63, 64–5
fraternity 162–3
free love movement 160, 162
freedom
Enlightenment thinking on 36
French Revolution 162
Friedman, Milton 39
friendship love 159, 161–3, 166

Gaia and the planetary commons 18
Galileo 36
gay rights 84
GDP (gross domestic product) 156
gender
Butler on gender identity 85–6
and fraternity 163
men and minoritarianism 87, 88
genetics
and microorganisms 51–2
genocide 82, 175
gentrification 152, 154
anti-gentrification campaigns 114–15
Germany 26
Berlin 44, 114
Nazi Germany 167, 168
The Ghan (television programme) 128–9, 130
Gibson-Graham, J. K. 5, 7–8, 20, 23, 186
global capitalism 1
global commons 8–9, 15, 18
global north
Mother Earth laws 75, 182
global south 178
global trade 37
globalization 18–19, 37, 49
Gnutella 118
goals, failing at 144
Google 114

211

Index

Governing the Commons (Ostrom) 15, 20, 153
Great Depression (1930s) 38
Greece
and the refugee crisis 50, 54
see also ancient Greece
greenhouse gas emissions 61
gross domestic product (GDP) 156
Growth Heathrow squat 71, 72–3, 74, 75, 153
Guattari, Félix 23, 112
concept of the body without organs (BwO) 53–4
Deleuze and Guattari's desire-production 121, 123, 126
and minoritarianism 81–2, 83–4, 89, 91, 94–5
Gudeman, S. 8, 9, 186

Haivan, Max 151, 174
Halberstam, Judith (Jack) 148–9, 157
Hall, Stuart 40
Hancock, Matt 97–8
happiness 124–8
and dopamine 124, 125
and serotonin 124–5, 125–6
UAE 'National Charter for Happiness' 124
The Happiness Industry (Davies) 124–5
Hardin, Garrett 15, 20
Hardt, Michael 16, 173–4
Hardy, G. H. 102–3
Harman, Graham 73
Harvey, David 153
Hayek, Friedrich 39–40, 41, 42, 133
healthcare practice and policy
and digital codification 105–6
Heatherwick, Thomas
the 'Vessel' 113
Heraclitus 11–12, 20, 45
'heroic' individualism 54–5
HFT (high-frequency trading) techniques 134–5

hippie subculture
'free love' movement 160
Hobbes, Thomas 36
homelessness 73, 188
and the coronavirus pandemic 26–7
Horvat, Srećko 160, 161, 167
Hudson Yards 'smart city' project 112–13, 114, 118
human body
capitalism and materiality 59, 60
and mutualism 51–3
and sensory augmentation 110–11
human trafficking 104
human/nonhuman worlds
and the coronavirus pandemic 185
and neomaterialism 60
and power 169
and transmaterialism 61, 72–3, 74–6, 123
see also animals
Hungary 29
Hurricane Katrina 60, 138
Hyper-Reality (film) 115–16
hyperhumanism 111

immanent ethics 22–3
imperialist mercantilism 1
India
Van Panchayats forest management 9–10
indigenous communities
and the commons 9, 11, 12–13, 15, 30
and failure 143–4
individualism
and capitalism 121
and failure 142
'heroic' 54–5
and mutualism 34–5, 45–7
and neoliberalism 39–44
self-interested 33–4, 35–9
Industrial Revolution 37

Index

inequality
  and failure 141
  and the family 165
  and minoritarianism 79–81
intellectual property rights 16
the 'interface envelope' 115–16
internet 18, 26
  see also social media
interpellation
  and the becoming-subject 92
intersectionality 84–9, 93–4, 174
'invisible hand' metaphor
  in Smith's *The Wealth of Nations* 36–7
IPCC (Intergovernmental Panel on Climate Change) 138–9
Israel 29, 84

Jameson, Fredric 77
Jesus 167, 168
Jews 168
Jobs, Steve 33
Johnson, Boris 184
justice 190
  power and love 169

Kaepernick, Colin 64
Kalanick, Travis 33
Katz, Cindi 94–5
Kelly, Kevin 105
Keynesian economics 38
Kiberan Revolution 182–3
King, Martin Luther Jr 168–9
Klein, Naomi 43
Kropotkin, Pyotr 45
Kurdi, Alan 50

laissez-faire economics
  and self-interest 37
land
  capitalist inequalities of 59
  primitive accumulation of 100
Latin America
  farm workers 64
Latour, Bruno 18, 118
legal codification 100
legal ethics 21–2

Leicester
  sweatshops 132
Leopold II, King of Belgium 82
the Levellers 13–14, 71
Lewis, Sara 143–4
Lewis, Sophie Anne 165
LGBTQ+ community 84, 145, 173
liberalism
  and self-interest 36–7
liberation
  and the growth of capitalism 122
Little Miss Sunshine (film) 146–9
Locke, John 36
lodestone 110
logos, the common of the 11–12
London
  South Bank skateboarding site 153–4, 155
Loraine, Tasmin 23
love 32, 158–76, 184, 190
  agapē 159, 166–76
  Bible (Corinthians) on 158–9
  eros 159–61, 167
  philia 159, 161–3
  storge 159, 164–6

marketization
  and codification 101, 106
  and fashion 133–4
  of higher education 165
Martin, Trayvon 86
Marx, Karl 121, 168
Marxism
  exchange value 99–100
  and materiality 57–60
materiality
  and *agapē* love 170
  neomaterialism 57–61
  see also transmaterialism
mathematics, fallibility of 102–5
Matsuda, Keiichi
  Hyper-Reality (film) 115–16
Mattern, Shannon 113
#MeToo movement 83, 88

213

Index

media
  mass media and solidarity 145
  slow media 123, 128–31
  see also social media
medical ethics 21
meditation 136, 137
men
  and minoritarianism 87, 88
mental health
  and the slowness ethic 137–8
  and social media 166
Mercator world map projection 83
Mercury, Freddie 103
Merry, Sally Engle 104–5
Mesopotamia 11
Mexico
  Cherán 10
the 'microbiome' 51–2
migrants 80, 83
mindfulness 137
minoritarianism 32, 77–96, 176, 184, 188–9, 190
  and *agapē* love 170
  becoming-minor 81–4, 85, 89, 91, 94–5
  blaming minority subjects 83
  and *Children of Men* (film) 78–9
  disability 89–92
  and failure 148, 149
  and Grow Heathrow 73
  intersectionality 84–9, 93–4
  and slow media 130
mirror neurons 47–9, 92
Mnuchin, Steven 163
monarchy
  and seventeenth-century ideas of the commons 13, 70, 71
Moore's Law 67
Moraru, Christian 18, 19, 21, 70, 131, 187
Moses 79
Mother Earth laws 75, 182
Mozilla Firefox 118
Muñoz, José 148, 149
museums 95

music
  slow music 129–30
Musk, Elon 33, 75, 183
mutual aid projects 127
mutualism 32, 33–56, 175, 184, 185, 190
  and commoning 46–7, 54–6
  and eco-squats 74–5
  and love 164, 170
  philosophical connections 53–6
  the science of we 47–53
  and self-interest 34–5, 44, 45–7, 56, 170
  and slow media 130
  and solidarity 144

NASA 181
natality
  and *agapē* love 171
Native Americans 12, 13
natural resources
  and the commons 7, 8
nature/culture divide 18
navigation, decodification of 109–11
Nazi Germany 167, 168
Negri, A. 16, 20
neoclassical economics 37, 39
neoliberalism 1, 17, 39–44, 49
  and *agapē* love 171
  and codification 99
  and the coronavirus pandemic 156–7
  and disability 90–1
  and eco-squats 153
  and failure 142, 143–4
  and the family 164–5
  and fashion 133–4
  and happiness 124–5
  and mutualism 54, 56
  and networks of political elites 162–3
  and the quantified self 106
  and the refugee crisis 49–50
  and smart cities 116
  and success 141

Index

Thatcherism 40–2
see also capitalism
neomaterialism 57–60
Netherlands
  digital commons platforms 116
The New Spirit of Capitalism (Boltanski and Chiapello) 120–3
New York City
  Amazon 'HQ2' 113–14
  Hudson Yards 'smart city' project 112–13, 114, 118
New Zealand 75, 156, 181
Newton, Sir Isaac 36
NGOs (nongovernmental organizations) 54
Nietzsche, Friedrich 168–9
numbers
  codification of 99–105
  and the coronavirus pandemic 97–9

objectivism 33–4
  and neoliberalism 43–4
objects
  commodification of 57–8, 67
  object-orientated philosophies 73–4
  right to repair 65–70
Ocasio-Cortez, Alexandria (PAOC) 179, 180, 182, 183
OECD (Organisation for Economic Co-operation and Development) 155
oil production 60
Orbán, Viktor 29, 184
Osborne, George 163
Ostrom, Elinor 15, 20, 153, 186
O'Sullivan, Michael 90
the Other
  and *agapē* love 171, 173
  and minoritarianism 82, 83
  and mutualism 49, 51

Paris
  1968 uprisings 25
  Commune (1871) 14, 20, 25, 150–2, 155
parliamentary democracy
  and the Levellers 71
Paul, Saint 167, 176
Peloponnesian War 35
Pence, Mike 179
PETA (People for the Ethical Treatment of Animals) 64–5
pharmaceutical companies 136
pharmacopornographic control 108–9, 127, 136
philia (friendship) 159, 161–3
Phillips, Henry 65–6
Piketty, Thomas 141, 165, 180
planetarity 19–21, 187
planetary commons 3, 6, 10, 17–21, 189–90
  and *agapē* love 175
  application of the seven ethics 177–84
  and art 189
  and capitalist neomateriality 60–1
  decodification 109, 111, 112, 116–19
  ethics of 30, 31–2, 184–5
  and failure 144, 146, 157
  and minoritarianism 79, 80–1, 83, 85, 87–8, 94, 96
  and mutualism 51
  and natality 171
  and the Paris Commune 151
  and the slowness ethic 123, 127, 128, 131, 134, 136, 137, 138–9
  and transmaterialism 74
pleasure
  and happiness 125, 126, 127
police
  and Black Lives Matter protesters 171–2
  and the coronavirus pandemic 29
politics
  love as a political force 173–4
The Politics of Friendship (Derrida) 162

Index

pollution
  AirCasting data-mapping platform 116–17
populism 28
  and neoliberalism 43, 44
  and objectivism 34
power
  Deleuze on codification 106–8
  and love 168–9, 174
  and success 141
PPE (personal protective equipment) 68–9, 97
Preciado, Paul B. 108–9
prejudices, structures of 88–9
Presciado, Paul B. 119
the present
  and the slowness ethic 134–9
primitive accumulation 100
privatization 4, 175
  and codification 106
  and the commons 15
  eco-squats 152, 153
  and urban decodification 115
production, slowing of 139
progress
  Enlightenment thinking on 36
The Protestant Ethic and the Spirit of Capitalism (Weber) 121
protests
  against smart city programmes 114–15
  Black Lives Matter 171–2
public transportation 118
Putin, Vladimir 181

the quantified self 105–9
quantitative/qualitative statistical analysis 104–5
The Queer Art of Failure (Halberstam) 148–9
queer studies 175
queering success 146–9

racism 79, 89
  and Black Lives Matter 80, 82–3
  reverse racism 87
rainforests 9

Rand, Ayn 33–4, 43–4
rationalism
  and self-interest 35–6
Reaganism 42, 43–4, 133, 164
Reclus, Elisée 14
recycling 67–8, 71, 180
  Grow Heathrow 73
refugees
  activist groups 5
  blaming 83
  crisis of 49–51, 54, 55
religion
  and *agapē* love 166–9, 173
  and the family 165
  and neoliberalism 44
  radical Protestantism and the commons 13–14, 70–1
'The Repair Foundation' 66
reparations
  from colonial powers to former colonies 95–6
research ethics 22
resilience 186
  and the slowness ethic 137, 138
resistance 190
resource depletion
  and the planetary commons 17
revolutionary love 167–75
revolutionary thought 149, 150–5
  and free love 160
Rifkin, Jeremy 48, 49
right to repair movement 65–70
The Road to Serfdom (Hayek) 39–40
Robertson, Peter 65
Ross, Kristen 150–1
Ross, Wilbur 163
Rousseau, Jean-Jacques 36
Roy, Arundhati 27
Russia 181
Russian Revolution 25
Ryan, Dr Mike 156

Said, Edward 161
Samsung 67
Sanger, Larry 119
scientific rationalism 82

Index

Scott, Alan C. 148
screwdrivers 65–6
self-actualization 169
self-care 55, 169–70, 174
self-fulfilment 144
self-help 186
   and failure 143–4
self-interest
   and the commons 14, 15
   and failure 142, 144
   history of 35–9
   and mutualism 34–5, 44, 45–7, 56, 170
   Randian philosophy of 33–4
   and the refugee crisis 49–50
   see also neoliberalism
sensory augmentation 110–11
serotonin 124, 125–6
sexism 79, 83
   reverse sexism 87
sexualized love (*eros*) 159–61
Sheely, Thad 112
Sidewalk Labs 114
Silicon Valley corporations
   and artificial intelligence (AI) systems 182–3
   failure and success in 140, 141
   protest against 114–15
Simon, Joshua
   Neomaterialism 57–8, 67
slave trade 37
slowness 32, 120–39, 176, 184, 186, 188, 190
   and *agapē* love 170
   and capitalist excitement 123
   and climate catastrophe 134–9
   happiness 124–8
   and the power of the present 134–9
   and self-care 169–70
   slow fashion 123, 132–4
   slow media 123, 128–31
   and the 'spirit' of capitalism 120–3
   and throwaway culture 123
   and well-being 136–7

smart cities 112–16
smart citizenship 115–16
smartphones 58, 67, 107, 111
Smith, Adam 36–7, 38–9
Smith, Stephen 90
social justice
   and mutualism 56
social media 16, 111, 134
   and dopamine 125
   and friendship 166
   photos and videos of Black Lives Matter 171–2
   and the refugee crisis 50
   and solidarity 144–5
socialism
   and the Paris Commune 150
solidarity 186
   and *agapē* love 174
   and failure 144–6
   and friendship (*philia*) 161
   working-class 188
Solnit, Rebecca 27
South Korea 29
Space X 181
Spain 26
   Barcelona 71
Spivak, Gayatri Chakravorty 3
squats 5, 71–6
SSASL (Summit of South American Socialist Leaders) 180–1
states, failed 155–7
statistical analysis
   reliability of 100–5
Stefan, Melanie 140
storge (familial love) 159, 164–6
success
   and failure 141–2, 146, 146–9, 154
Suffragettes 79
sweatshops 132

Taiwan 29
Tanner, Kathryn 134, 135, 136
Taoism 11
taxation
   'Robin Hood Tax' 180

Index

technology
 and capitalist development 120
 planned obsolescence 67
 see also digital technology
television, slow 128–9, 130
terrorist (9/11) attacks 25, 144, 145
Thatcherism 40–2, 133–4, 164
'there is no alternative' (TINA) 3, 42
Thiel, Peter 33
throwaway culture
 and slowness 123
 and transmaterialism 67–8, 123
Thucydides 35–6, 38, 44, 46, 82, 88
Tillerson, Rex 162–3
Tillich, Paul 169, 170
Tolstoy, Leo 168
trade union movements 188
'tragedy of the commons' 15
transmaterialism 32, 57–76, 176, 184, 186, 190
 eco-squats 70–6
 and homelessness 73, 188
 and neomaterialism 57–61
 right to repair movement 65–70
 and slow media 130
 veganism 61–5, 69
transphobia 83
travel
 and slow media 128–9
Trump, Donald 28–9, 33, 46–7, 98, 167, 179, 184
 hiring of political elites 162–3
twins 52

UAE (United Arab Emirates) 124
Uber 118, 183
UFE (United Federation of Europe) 181–2, 183
unconscious biases
 and disability 92
unemployment 38
United Arab Emirates (UAE) 124
United Kingdom
 coronavirus pandemic 97–9

 disability benefits scheme 89–90, 91
 FixMy Street 117
 government-led public health programmes 105–6
 Growth Heathrow squat 71, 72–3, 74, 75
 neoliberalism 41–2, 43
 political elites 162, 163
 Royal Mail 163
 UK government and the coronavirus pandemic 2, 26–7, 29
United Nations 8, 15, 186
United States
 AirCasting data-mapping platform 116–17
 coronavirus pandemic 98, 155
 as a 'failed state' 155
 future events (2020s) 179, 180
 and global trade 37
 Green New Deal 179, 180
 Hurricane Katrina 60, 138
 materiality 60
 the 'moral majority' 44
 neoliberalism 42–3, 44–5
 political elites 162–3
 terrorist (9/11) attacks 25, 144, 145
 urban decodification 112–14
 US government and the coronavirus pandemic 2
 see also New York City
universal basic income
 and the coronavirus pandemic 2, 26
universities
 academia and failure 140, 141–2
 league tables 100–1
 and the slowness ethic 137–8
 students and families 165
urban decodification 112–16
use value
 and slowness 127
USSR 44

Index

veganism 61–5, 69, 72, 74, 133, 188
Volkswagen 181
volunteers
  and the refugee crisis 50, 54, 55

Wales, Jimmy 119
Ward, Gail 90
Warren, Elizabeth 66
waste, global trade in 68, 69
Watson, Lilla 145, 174
The Wealth of Nations (Smith) 36–7, 38–9
wearable technology 105
Weber, Max 121
welfare state 38
well-being
  codification of 105–6
  and serotonin 126
  the wellness industry 136–7
WHO (World Health Organization) 156
Wikipedia 118–19
Winfrey, Oprah 143
Winstanley, Gerrard 13, 70–1
Wolf, Gary 105
Wollstonecraft, Mary 36
women
  and sexualized love 160
  and statistical knowledge on domestic abuse 104
Wood, Mark 90
working-class people
  and ethics 188–9
workplace
  cooperatives 181
  four-day working week 181–2
  well-being programmes 137–8
World Bank 8, 15
World Trade Organization 15
World War II 38, 69, 168
Worstward Ho (Beckett) 141
Wotton, Linda 89–90
Wren, Sir Christopher 36

yoga 136
Yusoff, Kathryn 59, 73, 74

Zimmerman, George 86
Žižek, Slavoj 62, 69, 77